19.95 8/80

APPLICATION DESIGN HANDBOOK For Distributed Systems

APPLICATION DESIGN HANDBOOK For Distributed Systems

ROBERT L. PATRICK

CBI Publishing Company, Inc.
51 Sleeper Street
Boston, Massachusetts 02210

Copyright © 1980 by Robert L. Patrick.
All rights reserved. No part of this book may be reproduced by any process or technique without express written permission.

Printed in the United States of America
Printing (last digit): 9 8 7 6 5 4 3 2 1

Library of Congress Cataloging in Publication Data
Patrick, Robert L. 1929-
  Application design handbook.

  Bibliography.
  Includes index.
  1. Electronic data processing. 2. System analysis. I. Title.
T57.5.P38   001.6'1   79-27205
ISBN 0-8236-1601-6

Compositor: UniScan Photographics, Glendale, CA

CONTENTS

Preface	vii
1. Application Economics	3
Introduction	3
System Life-cycle	5
Development Task Phasing	6
Staff Loading	10
Cash Flow	12
Investment Profile	14
2. Life-cycle Activities	18
Overview	18
Development Activities	20
Activity Highlights	26
3. Definition Phase	33
Project Management	35
Business System Requirements	44
Feasibility Study	50
System Analysis	56
Documentation	59
Physical Facility	60

4. Design Phase	62
Project Management	64
System Analysis	73
Design	102
Programming	193
Test Development	196
Documentation	200
Training	201
Conversion	203
Physical Facilities	205
Installation	206
5. Programming Phase	207
Project Management	209
Programming	210
Documentation	213
Conversion	215
6. Test Phase	218
Project Management	220
Test Development	221
Documentation	222
Training	223
Installation	223
Operation	224
System Maintenance	226
7. Operation Phase	228
Project Management	230
Operation	231
System Maintenance	234
Operations Management	236
8. Conclusion	239
Appendices	
A. Payout Worksheet	245
B. Sample Difference List	252
C. Bibliography	257
Index	261

PREFACE

The practicing systems analyst has long been slighted in the published computer literature. We have newsletters and magazines to keep managers informed about current events and pioneering applications. We have academic texts which record proven mathematical methods in excruciating detail. We have vendor literature which describes the available products in more or less understandable terms. But seldom does the practicing systems analyst get any help in unifying this information to produce a successful production computer application.

This handbook is a state-of-the-art manual. Its contents were distilled from books, periodicals, technical reports, consulting experiences, lectures, and interviews. When its contents were frozen (Spring 1979), it contained a distillation of the best current systems analysis practice as applied to small computer systems. Some of that practice was firm, well understood, and is so described. Other practices were tentative, preliminary, and appear to work in certain environments; these are described along with their limitations.

This is also a book of ideas, independent of any one vendor's machinery or software. It addresses distributed data processing applications which allow some of the processing to be performed and some of the data to be stored immediately adjacent to the originating source, with the remainder of the data stored and the remainder of the processing performed elsewhere. It makes appropriate distinction between peer coupled and hierarchical systems in a network, but it recognizes that many of the problems are common to both. It provides checklists for the analyst and follows the common time-phased system development process from beginning to end.

Since it is organized by phases, I recommend that the systems analyst first read this handbook from cover to cover to gain an appreciation of the topics discussed and the phases wherein they are important. Then as the analyst proceeds through the system development cycle on a new project, I recommend that the chapters on each phase be reread immediately prior to the initiation of that phase so the ideas contained in this handbook can have the greatest impact.

In the preparation of this book, many contributed. My peers, my clients, and my many work associates over the years each contributed indirectly to my idea file. Beyond that, Al Scherr and Hal Lorin contributed substance, advice, and criticism in healthy measure. I also wish to pause and apologize to the ladies. I attempted to write, and my editors attempted to produce, sexless prose. However, in some places the technical content and the eliptical phraseology of sexless sentences got too convoluted, so I resorted to the more direct and familiar masculine pronouns. No offense is intended to the many ladies among us.

One note of caution: Designing applications for a distributed environment is an emerging, rapidly changing art, not a science. The ideas presented here are the best available as this book goes to print. But if you have derived techniques superior for your environment to some of the ones contained here, do not hesitate to incorporate them, as that is the way the state-of-the-art is improved.

Robert L. Patrick
Northridge, California

**APPLICATION DESIGN
HANDBOOK For
Distributed Systems**

1
APPLICATION ECONOMICS

INTRODUCTION

Small computers have been manufactured since the early 1960s. For the first ten years of that period, they were almost exclusively used for dedicated purposes in either a stand-alone or integrated system environment. Most of these machines were purchased without software, and the application and software were both developed concurrently by the purchaser. In the last eight or ten years, some manufacturers have developed packaged applications for specific markets. Where the design specification for the application package matches the customer's needs, these have been successfully installed and operated. However, these packages have tended to address very specific applications areas and have been only slightly adaptable to many customers' needs.

In recent years a few manufacturers have started to offer small computers with enough configuration flexibility and sufficient software to allow them to be used for general purpose commercial data processing. Manufacturers now provide a wide variety of input/output devices and basic software which consists of the operating system, compilers, utility programs, and libraries traditionally supplied by the vendors of large equipment. Recent announcements have driven down the prices on these offerings to record lows.

Current small equipment offerings can be successfully used for general purpose data processing, data capture, and remote job entry. These features plus the host support packages that are available from the major mainframe vendors, allow the small systems to operate stand-alone, as a processor in a peer to peer network, or as an intelligent node in a hierarchy with a larger host mainframe.

The new crop of small computers provides a wide range of communications capability, reliable hard disks for the storage of remote data, and some degree of compatibility so the application designer can prepare programs even if part of the processing and data storage reside at the remote site nearest the users, and the remainder of the data and processing resides elsewhere.

Experience indicates there are some sharp and significant differences between developing applications for a small computer and developing applications for a large host system. The onset of performance and configuration problems is more abrupt in the small system, and fewer hardware options are available for the solution of those performance problems. If a set of related application programs were to outgrow the installed configuration, off-loading some of that application processing on to a second small computer is sometimes difficult unless that transition was considered when the application set was initially designed.

Inexpensive small computers allow data capture and at least some fraction of the subsequent processing to be performed near the source of the data. Further, distributed files allow summary data to be readily available for user inquiry. The application designer should realize that while distributed processing and distributed terminals provide processing to a broader population of users, they can present new problems unless the design is carefully contructed, and both human factors and operational aspects are considered.

If the application designer is successful in providing an interactive system with storage and processing where it is needed most, the system will ingratiate itself to its users, and they will rapidly become dependent upon it. <u>In contrast with batch systems which fail privately, on-line systems fail publicly and impact their users immediately.</u> Although equipment manufacturers have provided the most reliable systems possible using today's hardware and software technology, failures will occur. Applications programs which do not acknowledge this possibility will fail hard, be difficult to restart, and will be unacceptable to their population of users. This manual attempts to assist the designer of application programs who is thoroughly familiar with the host systems environment, yet is designing a first application for a small computer. This manual does not attempt to be a primer and teach applications design per se, but it does try to highlight the differences between the host and the small computer environments.

SYSTEM LIFE CYCLE

Before proceeding with a lengthy enumeration of design hints, it is appropriate to step back and look at the overall process. A life-cycle approach provides needed insight to developers who have spent the majority of their professional life in technical pursuits and have not been too concerned with the managerial aspects of data processing. At the risk of stating the obvious, remember that the development phase of a system usually lasts a few months while the operational phase of a system lasts a few years. Recent cost analyses have shown that the cost of operations and maintenance is usually several times the development cost. Therefore it seems appropriate to consider the entire life-cycle of an application — a period that starts with a user's stated need and ends when the operational system has been discarded due to technological obsolescence.

DEVELOPMENT TASK PHASING

Figure 1.1 depicts the phasing of development tasks and the initiation of production operation for a typical small computer system. The figure lists 15 tasks which must be scheduled, staffed, and budgeted during the planning phase of the project:

1. Project Management
2. Business System Requirements
3. Feasibility Study
4. Systems Analysis
5. Design
6. Programming
7. Test Development
8. Documentation
9. Training
10. Conversion
11. Physical Facilities
12. Installation
13. Operation
14. System Maintenance
15. Operations Management

As shown, the collection of tasks and their phasing is typical of the applications development process in a small machine environment. (The process depicted is for illustration only and should not be used in the absolute sense to judge actual development plans.) It typifies the tasks necessary to automate an existing manual system and put some users on-line. If the manual system did not exist and invention was required, the Requirements, Feasibility, and Analysis tasks would be greatly expanded.

If the processing to be accomplished challenges the capacity of the hardware to be installed, then the Design task would be enormously increased, with the possibility that the application could not be made to fit the chosen configuration at all. On the other hand, Training and Conversion might be overstated in

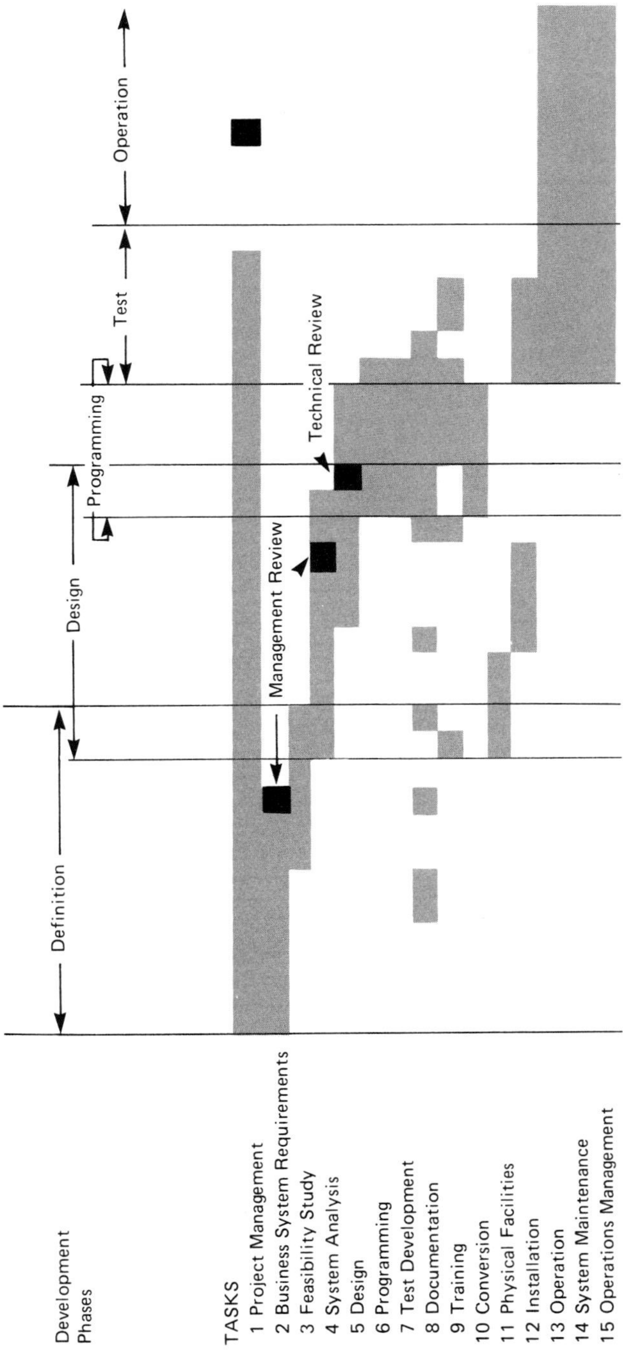

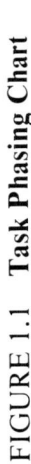

FIGURE 1.1 Task Phasing Chart

Figure 1.1 if the designer was fortunate enough to be working with experienced users and if the data files were already validated and in machine-readable form. Further, the Installation task might be greatly simplified if the small computer was being installed in an existing facility. Thus while the tasks depicted are typical, a great variety exists even with a simple stand-alone installation of medium complexity.

Before proceeding, several points of interest should be noted in the task phasing chart shown in Figure 1.1. The chart depicts the project management process as continuous from the beginning of the definition phase through the end of system test. The small isolated block of management activities that occurs during the Operations phase depicts a post-installation audit.

Managements frequently request formal status meetings so they can be apprised of the project and its progress. Two of these are shown: one when the requirements are completed, and one when the systems analysis is well in hand. Locating the meetings at these points provides management with the maximum leverage over the developing system. First, assurance is provided that the requirements are in consonance with the business, and second, development and operational costs are confirmed after the preliminary design has been completed and the designer is in a position to revise earlier estimates.

The chart is an attempt to depict a slightly idealized real-world situation. Documentation takes place in parallel with the other tasks, and physical facilities are completed just when required for equipment installation, which in turn is completed just when the equipment is required for initial program test.

Of particular note is the overlap between the Definition, Design, and Programming phases. Once a project has been initiated, managers are usually inclined to accept a little risk and compress the schedule. The overlap between Definition, Design, and Programming cuts about 10% off the schedule. If the definition of the work to be accomplished is well understood at the beginning of the project, further overlap would be possible. On the other hand, if the design is absolutely critical to the system's success, it might be impossible to start any programming before the final design is in hand and thoroughly reviewed.

A popular myth states that applications on small computers are simple, easy, straightforward, and quick; whereas applications on host systems are just the opposite. However, experience has shown that any application on a dedicated system tends to be easier than the same application in a shared environment. Well-defined simple applications that do not tax the installed hardware are in fact easy to install anywhere, but undefined systems require considerable analysis prior to programming. Complex applications are innately difficult regardless of whether they are installed on a small machine or in a large host environment.

Many experienced project managers find charts such as Figure 1.1 insufficient for depicting the complexity of their development activities. Instead they prepare some form of simplified network chart (a la PERT) which depicts both the activities that must be conducted to accomplish each task and the dependencies between those activities. They contend that dependency charts allow project planning to be conducted at a primitive level based on atomic units of work. After planning at this fundamental level, they then aggregate their plans and produce bar charts such as Figure 1.1.

In any event, the next step in project planning requires each unit of work to be identified and an estimate of the labor and other resources required for its accomplishment to be determined. During this estimating process, individual labor categories must be specified so that the skilled personnel required for each activity are available when required. Since the cost of labor usually varies with the skills, a project leader will usually contact an administrator and obtain labor rates to be used for estimating purposes.

A project involving a small computer can range in size from a simple stand-alone installation involving one computer and only local loop communications, up to a network of small computers, each with its own local loop communications, and each connected by peer-to-peer or hierarchical links to adjacent processors. Thus the personnel and skills needed cover a wide range. Specifically a given project can involve any or all of the following:

    Project Manager
    Project Administrator
    Chief Designer and Assistants
    Chief Analyst and Assistants
    Development Programmers

> Data Base Administrator
> Communications Administrator
> Central Support Staff
> User Management
> Users/Operators
> Remote Site Manager
> Central Operations Manager
> Maintenance Programmers
> Vendor Support Personnel
> Vendor Maintenance Personnel

When a designer sets out to develop an application system, he must be familiar with the roles played by all of the personnel just listed. In some ways staffing for a large system design is easier than for a small one. In a large system the designer can identify specific individuals with each of the above responsibilities, interview them, and in some cases get them to sign off on critical aspects of the system design. However, in small systems, one person may play many roles. The local site manager may be the user manager and the lead operator. Thus the designer of a small system must think through all of the aspects of an emerging design before it is considered complete. Remember: *Design is an iterative process.*

Since some aspects of the design are coupled, a positive tradeoff on one aspect may provide negative benefits elsewhere. Thus each design is a compromise, and the designer must recognize these tradeoffs as they occur in the design process.

STAFF LOADING

As discussed earlier, the project leader can itemize all the activities to be performed in support of each task and estimate the effort required from each. If this were done for a typical project, the project staffing could be superimposed on the task phasing with results similar to Figure 1.2. This Staff Loading chart shows the manpower buildup corresponding to the Task Phasing discussed earlier. Note the relationship between the phases and the staff buildup. The first management review occurs at the end of the Requirements task and precedes the staff buildup and expenses related to the Design phase. The second management

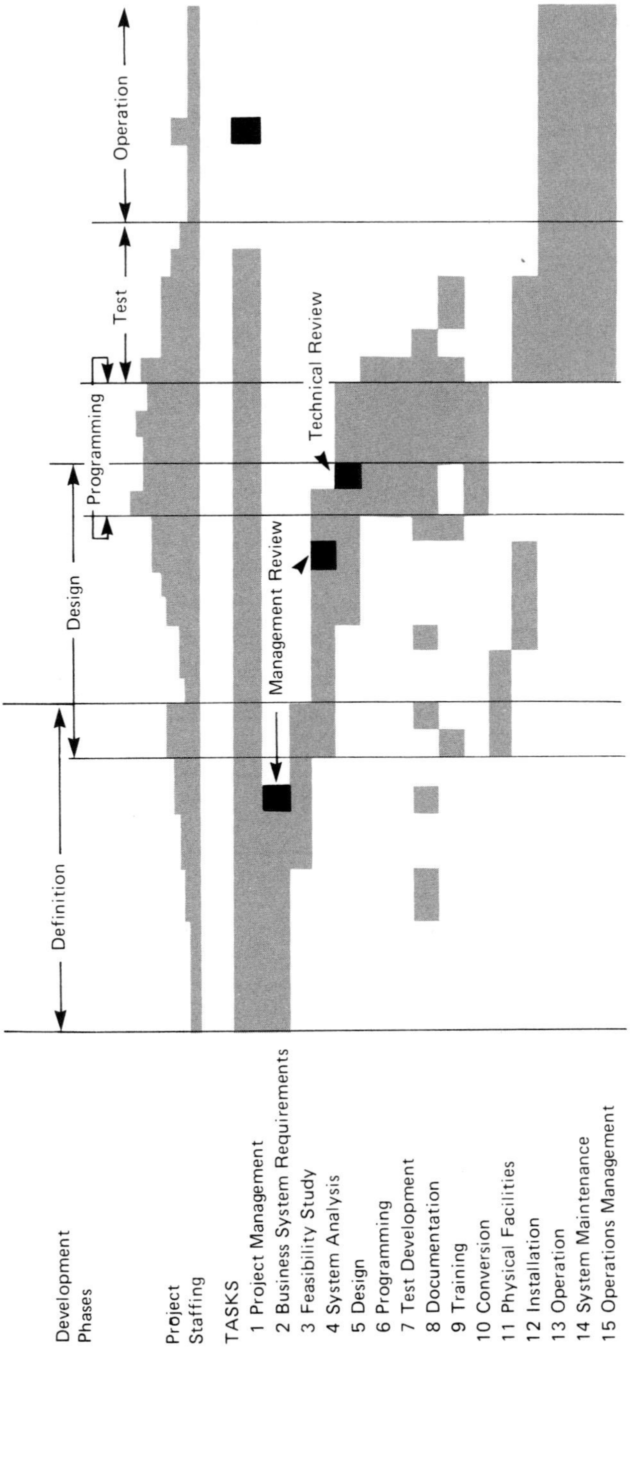

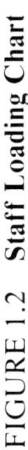

FIGURE 1.2 **Staff Loading Chart**

review occurs towards the end of the Analysis phase and precedes the staff buildup for the Programming phase. The technical review shown at the end of the Design task is a go- no-go decision point. A peer review by another set of designers will allow design flaws to be caught before they are encased in programs, and in the rare case that a designer is confronted by a task beyond his abilities, will allow additional skill to be injected into the project before irrevocable decisions are made.

After the tasks and activities are enumerated and after a technically qualified manager has estimated the resources required for each, an administrator can prepare a project cost projection. Appendix A contains an Applications Development Financial Payout Worksheet. Column 1 contains the project schedule. The entire system life-cycle is depicted in weeks from project initiation through replacement of the production system some years later. Columns 2 through 7 show the expense of the project, first in the development phases and later throughout its production life. Columns 8 through 13 depict the savings brought about by the system. During the development phase the savings are nil, but after production operation is initiated the savings begin to accumulate and usually improve until the system is replaced. Column 14 is the net cost/savings for the project.

This worksheet is the beginning of a Return On Investment (ROI) analysis. It should be noted that the need for systems is frequently justified, at least partially, by intangibles such as more data for decision making, the substitution of automated techniques for manual skills that are in short supply, faster response, more accurate processing, and more consistent processing. However, even though the necessity of a system may be partially, or even totally, justified on these intangible considerations, an ROI analysis allows the project manager to justify the system on hard tangible returns or to discuss intelligently the dollar savings shortfall so it can be determined if the intangible returns are worth the difference.

CASH FLOW

If a Payout Worksheet is completed, and if the net expense is superimposed on the task and staffing charts presented earlier, a Development Cash Flow Chart similar to Figure 1.3 results.

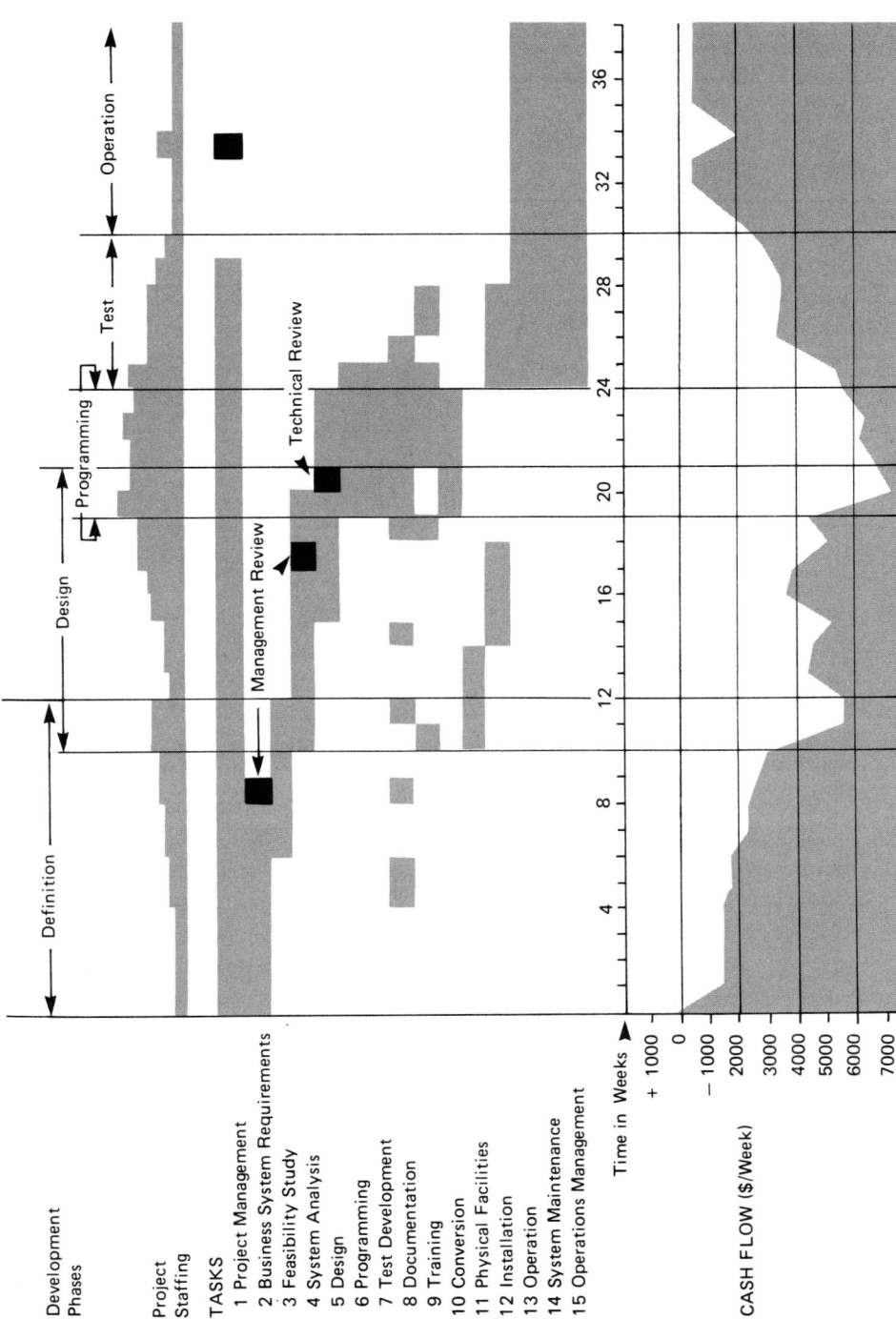

FIGURE 1.3 Development Cash Flow

Note how the cash flow and the development phases coincide. The first management review occurs after you know what is required but before the heavy expense occurs. The second management review and the technical review similarly occur at critical times in the expense profile. In week 45 the system has been in production long enough to begin accumulating positive tangible benefits, and the cash flow has turned positive. At this point the rate of expense is once again equal to the rate of expense prior to the initiation of the project.

A few notes about the costs presented are in order. First, the wage scales used were those prevalent on the U.S. west coast in the Fall of 1978. Second, the computer hardware was installed in week 14. This chart assumes current accounting practice such that the hardware was capitalized, and software lease, depreciation, and maintenance charges were incurred from week 18 forward.

INVESTMENT PROFILE

If columns 1 and 14 from the Payout Worksheet are plotted, an Investment Profile results similar to Figure 1.4. At project initiation there is expense and no compensating savings, so a loss is shown. This loss continues to accumulate until week 45 when the expense has been reduced and the saving has been built up so a net saving results. This is shown on Figure 1.4 as the point of maximum investment. If the saving continues to accumulate as planned, the saving will eventually offset the accumulated expense and the cash flow will remain positive until the system is retired. With the assumptions made, this crossover occurs in the 163rd week. In week 45 a sunk cost of almost $127,000 (plus the capital cost of the hardware) has been made. In week 163 all of that investment has been earned back by the accumulated savings, and if the system were to be used through week 249, a net $100,000 profit would result. The magnitude of these numbers demonstrates why managements require assurance that computer development dollars are well spent.

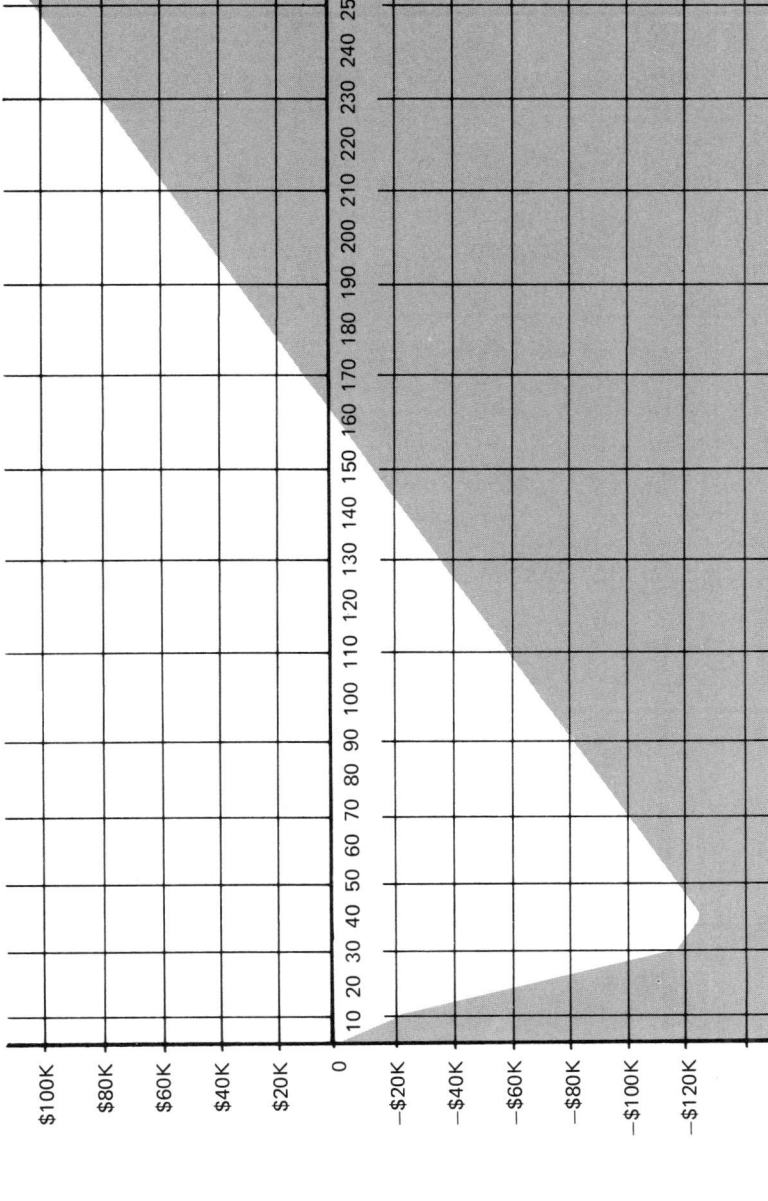

FIGURE 1.4 Investment Profile
(Calendar Weeks since Project Inception)

Now consider the problems that can occur if a development project is not properly undertaken. If the project is not properly organized, if the available skills are not up to the tasks, or if additional user requirements become visible during testing, then the development will be protracted. As a result the cash flow depicted on Figure 1.3 will cross the axis later and if the situation were bad enough, the investment curve on Figure 1.4 might not get into the black before the system was retired from service.

If a system were installed stand-alone with absolutely no thought of communication with an adjacent system, and later events such as corporate merger, new plant construction, or unforeseen business growth required communications, then considerable rework would be required to split a data base and hand off transactions to a peer. If this were to occur, a redevelopment project would be initiated part way through the system's life, requiring more analysis, design, test, retraining, and revisions to the production documentation. This redevelopment effort would have a major impact on the investment profile depicted in Figure 1.4. While it could be rationalized that these were unforeseen circumstances not under the purview of the original project leader, careful consideration during the initial design would keep the added costs to a minimum and would reduce the impact of the second system on the first.

Over the years many application designs have modified standard software and integrated application code and software into one monolithic package. While such design techniques are clearly justified in some cases, the increasing price of labor and the decreasing price of hardware are reducing the justification for such integration.

This is not to imply that the standard software should not be extended if a configuration involves unsupported I/O devices or unique application environments. However, even in these cases, the designer should maintain the established interface between applications code and software. Whenever software extensions are required, these should conform to the same standards used by the software developers so they do not unduly hamper application of software fixes issued by the vendor. To do otherwise will impact the investment profile shown on Figure 1.4.

Thus an application should be designed so it can grow without major reprogramming, so stand-alone systems can be coupled into a network without extensive rework, and so workloads can be split whenever the capacity of a single processor is exceeded. To plan for these requirements, the designer must be a visionary. He must think through the problems that are likely to occur, provide flexibility at the appropriate points, consider the system from the vantage points of the various persons who will come in contact with it, and produce a system that performs the desired functions within a schedule and a budget. It is axiomatic that the project will have to be done over if it is not done right the first time. Thus each designer must negotiate with his management for enough time and enough resources to prepare an adequate design.

The remainder of this manual enumerates some of the items a designer must consider.

2
LIFE-CYCLE ACTIVITIES

OVERVIEW

A small computing system can be used in a variety of ways. Even in a stand-alone mode, a degree of complexity is introduced when the system supports multiple applications. Another degree of complexity is introduced if each of the multiple applications has a different sponsor. If a small computer serves only one master (cost accounting, for example), operational priorities and training problems are rather easily resolved. If a system serves multiple masters, problems of operating schedules, space allocation, and system control are amplified. If the small system has only local loop communications, problems are contained and easily solved. But as soon as the system is connected to a common carrier, the problems blossom, and if a small system is a node in a network, either peer-to-peer or hierarchical, the simple applications environment, frequently discussed by authors in the popular press, has just disappeared.

Complexity is a matter to be considered warily by the designer. Underestimating the level of complexity may cause decisions to be made which cannot be undone without expending significant time and dollars for rework.

System complexity is dominated by network complexity. Consider the ascending complexities in the following five environments:

1. Stand-alone system, all units centralized.
2. Stand-alone system, with terminals on a local loop.
3. Stand-alone system, with remote terminals.
4. Two or more processors directly connected by communication links.
5. Multiple processors in a mesh network with alternate paths.

The system complexity is multiplied by the applications complexity, which in turn depends upon:

A. The data base configuration.
   1. Centralized.
   2. Partitioned.
   3. 100% redundant.
   4. Partially redundant.
B. Program configuration.
   1. Discrete independent applications run locally.
   2. Interrelated applications run locally.
   3. Applications which run to completion locally and then pass information on to the network.
   4. Applications which require simultaneous coordinated processing at two or more nodes of the network.

Further, the system complexity can be increased by the environment in which the application system is embedded. Consider the following factors which can make an environment challenging to the designer:

A. The number of intervening communication links and neutral nodes between the processor where a transaction is originated and the processor where the final file update occurs.

B. The variety of speeds and protocols in the communication links.
C. The variety of communication equipment types, makes, and models connected to a single system.
D. A requirement for dynamic allocation of communications, processes, and data sets.
E. Major variations in workload which can occur hourly, daily, or long term.
F. The variety of services offered in a single network such as batch, RJE, time-sharing, or interactive.
G. The requirement to routinely intermix data entry, inquiry, production, and system development.
H. Critical performance or availability requirements setting limits on response time, allowable blackouts due to an outage, or stipulating no-break operation.
I. The variety of equipment technologies and hardware/software design standards represented, whether from a single vendor or from multiple vendors.

Fortunately some of that system complexity can be reduced by proper management. However, even if the application system is properly designed, it is unlikely to be a complete success unless the following are also addressed:

A. Organizational roles and missions.
B. Training of all personnel.
C. Applications procedures (both manual and automated).
D. Systems operations (routine and abnormal).
E. Tools for problem determination and diagnosis.
F. Planned reconfiguration options.
G. Traffic and service level measurements.

DEVELOPMENT ACTIVITIES

Obviously the applications designer must know his application and his environment equally well. This is more true in the small systems environment, where a single application tends to dominate the configuration, than in the large host environment where the configuration of hardware and software is virtually limitless.

Figure 2.1 enumerates the specific activities which are most usually associated with the tasks discussed in Chapter 1. An attempt has been made to pinpoint the usual activities that take place throughout the life-cycle of a system. Ninety-six activities are enumerated and each of these is related to one of the 15 tasks discussed previously. In addition, the right hand portion of Figure 2.1 contains a matrix which shows the relationship of Task/Activity combinations to project phases as defined in Chapter 1. While most task/activity combinations fall into a single development phase, there are some, denoted by multiple Xs in a single row, that apply to two or more development phases.

While a brief reading of the 96 activities may be in order at this point, this list should be consulted whenever a development project is being planned; it will serve as a useful list for checking the adequacy of the planning process. Further, the astute project leader will add activities to the ones given here to build a customized checklist uniquely suited to his particular development environment. With 96 items on the initial checklist, some project leaders understandably elect to prepare dependency charts so the interrelations between activities can be understood and scheduled.

As was stated earlier, design is an iterative process. While the activities and tasks are time-phased as shown in Figure 1.1, the designer must think through the implications of all 96 activities when he is building a plan, going after requirements, preparing a preliminary design, and constructing a detailed design. If a designer were to overlook one or more of the activities, he might prepare a design which would function well in an ideal environment but could not recover from the errors encountered in a real environment. Or he might construct a design which assumed a level of skill and understanding beyond the experience of the user's terminal operators. Many computer applications that have been designed without built-in statistics and measurements eventually suffer from deteriorating performance that cannot be diagnosed.

22 Life-cycle Activities

			Applicable Phases				
			DEF	DES	PGM	TES	OPN
1 Project Management	A—	Plan Review	X				
	B—	Progress Tracking	X				
	C—	Financial Controls	X				
	D—	Staffing and Talent	X				
	E—	Milestone Reporting	X				
	F—	Change Control		X			
	G—	Post Installation Audit					X
	H—	Standards	X	X	X	X	
	I—	Development Methods	X	X	X	X	
2 Business Systems Requirements	A—	Functional Goals and Objectives	X	X			
	B—	Development and Operations Cost Targets	X	X			
	C—	Schedule Constraints	X				
3 Feasibility Study	A—	Technology vs. Requirements	X				
	B—	Project Plan	X				
	C—	Preliminary Estimates	X				
4 System Analysis	A—	Data Gathering	X				
	B—	Document Current Processes	X				
	C—	Preliminary Design		X			
	D—	Preliminary Cost Justification		X			
5 Design	A—	Environment		X			
	B—	Configuration		X			
	C—	System Aspects		X			
	D—	Man-machine Dialog		X			

Applicable Phases

	DEF	DES	PGM	TES	OPN
5 Design					
E— Data Bank		X			
F— Communications & Network Aspects		X			
G— Processing		X			
H— Procedures		X			
I— Controls		X			
J— Training		X			
K— Performance		X			
L— Testing		X			
M— Conversion		X			
N— Restart/Recovery/Availability		X			
6 Programming					
A— Decomposition			X		
B— Techniques			X		
C— Data Description			X		
D— Input			X		
E— Output			X		
F— Processing			X		
G— Unit Test			X		
7 Test Development					
A— Tools		X			
B— Integration		X		X	
C— Abnormal Conditions		X		X	
D— Acceptance		X		X	
E— Performance				X	
8 Documentation					
A— Requirements	X				
B— Specifications		X			

FIGURE 2.1 **System Life-cycle Activities**

24 Life-cycle Activities

			Applicable Phases				
			DEF	DES	PGM	TES	OPN
8 Documentation							
	C—	System Overviews	X				
	D—	User	X				
	E—	Operator			X		
	F—	Maintenance					X
9 Training							
	A—	Programmers	X	X			
	B—	User Management	X	X			
	C—	User Staff					X
	D—	Central Staff					X
	E—	Central Operations					X
	F—	Program Maintenance					X
	G—	Problem Determination					X
10 Conversion							
	A—	Transition Planning	X				
	B—	File Cleanup		X	X		
	C—	Standing File Conversion		X	X		
	D—	Work In Process Conversion			X	X	
	E—	Controls and Audit Trails			X		
11 Physical Facilities							
	A—	Space	X				
	B—	Power	X	X			
	C—	Air Conditioning	X	X			
12 Installation							
	A—	Equipment	X	X			
	B—	Preliminary Acceptance	X	X			
	C—	Applications Programs		X		X	
	D—	Convert Files				X	X

Development Activities   25

		Applicable Phases				
		DEF	DES	PGM	TES	OPN
12 Installation	E— Hands-on Training				X	
	F— Revise Processes				X	
	G— Parallel Runs				X	
	H— Cutover				X	
13 Operation	A— System Management				X	
	B— Procedures				X	
	C— Production				X	
	D— Problem Determination					X
	E— Operational Reports				X	
	F— Data Administration				X	
14 System Maintenance	A— Capacity Planning					X
	B— Program Fix				X	
	C— Program Extension					X
	D— Program Improvement					X
	E— Measurement and Tuning					X
	F— System Reconfiguration					X
15 Operations Management	A— Service Measures					X
	B— Productivity Measures					X
	C— Workload Projection					X
	D— Saturation Prediction					X
	E— Analysis of Improvements					X
	F— Justification of Upgrades					X
	G— Technology Trackign					X
	H— Continued Training					X

FIGURE 2.1   **System Life-cycle Activities (Continued)**

26   Life-cycle Activities

ACTIVITY HIGHLIGHTS

A quick trip through Figure 2.1 will highlight those items which are unique to a small computer environment or require different emphasis when compared with a large host environment. Under Task 1, Project Management, activity 1.F concerns itself with change control. In a large host shop the responsibility for change control usually lies with an independent production control section. Further, a modern shop will have separate development and production libraries with a formal procedure established enabling a program module to migrate from one to the other. In a small shop these matters must be considered and appropriate solutions defined. In a network where programs can be downline loaded, it must be possible to know what versions of programs are installed where, and the proper degree of control must be established so incompatible versions are not enabled for production operation simultaneously. (Some vendors supply network software to provide these control services.)

The remainder of Task 1 addresses the modern development methods that should be defined early by the project manager for the benefit of the entire project. Note that even small projects consisting of only one or two people benefit from this same discipline. Defining standards in writing is helpful even with only two workers, as it guarantees consistency and provides a program base which reduces the cost of maintenance.

Task 2, Business Systems Requirements, addresses a series of activities that have become increasingly important in recent years. Twenty or even ten years ago, the benefits of early automation were so dramatic that justification was underplayed so developers could proceed with the work of creating the application systems. However, in recent years, most of the applications with overwhelming benefits have been automated, and managements have become more cautious. Much recent trouble has been traced to incomplete and misleading requirements; requirements which did not differentiate between needs, desires, and wishes; or requirements which failed to stipulate operational schedules, target operating costs, or performance norms. Several analysis methodologies are useful for the determination of business systems requirements. Some of the more useful ones are listed in the Bibliography.

Unless the need for a system is so overwhelming that it must be designed and installed at all costs, Task 3, Feasibility Study, is the next logical step. When the feasibility study is completed, the project leader will have a good idea what hardware and software he will use, as well as the magnitude of the applications system development effort. He will have taken an initial survey of resource availability and will have devised preliminary cost and schedule estimates. At this point management can determine if the benefits (tangible and intangible) outweigh the estimated costs, and if the operational date yielded by the natural development schedule is satisfactory or if the project needs to be replanned and recosted to compress the schedule.

Task 4 discusses Systems Analysis. The series of activities itemized is based on the assumption, as is common, that the application already exists and is being performed manually or through semiautomated techniques. Proceeding on this basis, this task provides for documenting the present process, determining data flows and volumes, and enumerating error experience and response time. Frequently designers of new systems have ignored the present system and when the new system was delivered, were surprised to find that it was inferior in some respects to its predecessor. Even though the new system may have been better overall, slight deficiencies have detracted from its appeal and slowed its acceptance by users. A good systems analysis will produce cost and performance targets for the new system so it will either exceed the old system in every way or will have deficiencies known to the designers in advance of implementation.

Task 5 covers all aspects of Design. While this is the main thrust and emphasis of this manual, several activities are still worth highlighting here. First, the application running on a small computer in a remote location tends to dominate the organization in which it is imbedded. Thus the designer must consider the manual work processes in that environment, the movement of paper and material, and the user manager's need to be informed about the system's operation and how it is being used.

While the environment of a host system tends to be rigid, the environment of a small system is flexible and must be designed as part of the application design process. The human factors aspect of distributed systems design was mentioned earlier and is also involved in activity 5.D, Man-Machine Dialog. All designers building transaction-oriented systems should be familiar with at least one of the human factors references given in the Bibliography.

Activity 5.I reminds the designer of a distributed system that controls must be conceived and built into the system at its inception. Dollar controls, item counts, storage space, occupancy levels, error counts must all be built into the application code. In a large batch shop these controls are usually part of the application, and the control reports are concatenated with the normal output stream so they are routinely printed and delivered back to the production control group. In a distributed system the presence of a production control group at each location is not guaranteed. If the same application is running on two or more processors, some provision must be made for centralized reporting of activity statistics and for summarizing them at the site of systems maintenance.

Activity 5.K, Performance, is listed because small computers tend to saturate more abruptly than do larger machines. This is probably due to the tendency of one application to dominate the processing capacity in a small machine, whereas a large host machine running thousands of jobs a day benefits from the laws of large numbers. Thus performance must be continuously considered by an applications designer. Initially counts of transactions, statements to be executed, disk records to be accessed, and lines to be printed will provide a preliminary view of applications performance. Later the number of screens and the volume of data transmitted to and from each display will need to be calculated along with instruction path lengths, communication transmissions delays, and disk loadings. If performance will be a serious problem on your small system, additional work will be necessary. Good performance prediction techniques are not yet common practice. Good references are also scarce. If you have a critical problem, design a benchmark and run it on a real system or acquire the services of a consulting specialist.

Activity 5.M reminds the designer that a serious conversion problem may result if he is automating an existing system. If an application exists on the host and part of that application is being migrated outboard, then a major effort to convert and cleanse files of manual data may not be required. However, even if a machine readable file exists, the data may not possess the appropriate quality and integrity to be used by the new application. Or the new application may offer new features which are predicated on data elements that do not appear in the existing data set. In some cases the file conversion and cleanup activities have required more code and taken more labor than developing the code for routine application processing.

Activity 5.N reminds the designer that users rapidly become vitally dependent on on-line systems and hence become extremely concerned if the system is unavailable and their displays are blacked out for any significant portion of their working day. The best way to guarantee that small perturbations will not disrupt an entire system for an inordinate period of time is to consider the system aspects early in the design cycle.

Task 6, Programming, contains Activity A, Decomposition. After the preliminary design and the detailed design are completed, programmers still must know exactly how to construct records from data elements, where to place data, how to package processing functions, and where to place the control action responsibility if counts do not balance. The Decomposition must be conducted by a work party consisting of programmers and designers together. Here the aspects of growth, functional migration, and future networking must be exposed. Flexibility for future change without extensive rework is accomplished at this point.

Task 7, Test Development, contains two items which may not be familiar to designers experienced only on large host systems. A big host has a full set of test tools, performance analysis aids, and tuning techniques provided by the system programming department. In a distributed environment, some of these responsibilities fall to the applications team. It is anticipated that a large shop would have a section within the systems programming department devoted to supporting applications programming teams with the proper tools and techniques. However, the application teams would still need to define their needs and anticipate their requests even if the central systems support department were supplying the necessary tools. In a small shop the applications team is all there is, so the designers must be aware of the readily available tools for testing and tuning and must consciously decide whether the available tools are adequate or whether additional tools and techniques should be produced or purchased.

Tasks 8 and 9 address Documentation and Training. The activities enumerated here are straightforward, once the designer has clearly in mind the responsibilities to be assigned to each of the interested parties. If the remote nodes are small and if no professional data processing people reside at these sites, then the users and their management must receive broader training than has been traditionally required. Thus activity 8.C, Systems Overviews, is required to give the user a general understanding of the system. With stand-alone systems, some of the training activities may not be required. However with networks, application and operational responsibilities must be properly assigned and personnel trained to levels appropriate to those responsibilities.

Activity 9.G, Problem Determination, requires new emphasis. Intelligent networks can become complex, and since users are likely to be intolerant of system instability, problem determination and diagnosis responsibilities must be distributed along with the processing power.

If your small computer is connected to a host which contains a programmed maintenance and support facility, system support can be organized so a central support group can be located at the host and can remotely perform nearly all the maintenance, service, and control functions necessary for problem determination, problem isolation, and remote system control.

Task 10 elaborates on File and Data Set Conversion activities. It recognizes that transitioning an existing system from an old production support base to a new production base requires planning. Even if the bulk of the data is converted in advance, one still must deal with the work in process, and these are sometimes the most vital records. Further, as controls and audit trails have become more common in nonfinancial systems, internal auditors have become less sympathetic to conversion efforts which do not provide audit trail continuity.

Tasks 11 and 12 deal with Physical Facilities and the Installation of equipment and programs. A large shop has equipment planners who routinely worry about the activities listed. While most small computers are less sensitive to their physical environment than some pieces of host equipment, these aspects of physical planning cannot be ignored without the risk of uncertain communications, lack of physical security, electrical power transients, or abnormal heat and humidity which can affect the system's availability. These installation activities constitute a fairly straightforward sequence for any project leader who has been installing large on-line data base applications. Designers not possessing these skills should set aside some time and effort in their project plans or their applications may not install smoothly.

Task 13, Operations, addresses those activities which are performed automatically and routinely in a big host shop and constitute part of the responsibilities to be migrated outboard with the processing. In a remote facility, these six activities must be performed remotely. Part of the appeal of a small computer is the control the user and his management get over the day-to-day operation. However, when they assume that degree of control, they need operational reports, and they assume part of the responsibility for problem determination and data administration.

Task 14, System Maintenance, contains the usual activities of program fix, extension, and improvement. However, in a distributed environment, the maintenance crew must also perform capacity planning, measurement and tuning, and system reconfiguration, activities that are normally performed by the technical operations staff attached to the host system support group.

Task 15 addresses those Operational Management activities which while performed in a large central facility, assume an increased importance and an increased complexity in a distributed environment. Service and productivity are usually measured by the staff of the DP Director. Sometimes these measures are informal, and sometimes raw machine usage data is distilled to produce turnaround time statistics and counts of abnormal job terminations. In a distributed system the users are also concerned about keyboard response and availability. If the personnel maintaining the systems are not physically located in the same building as the remote system, they will have difficulty interpreting reports of service level deficiencies unless some service data capture has been built into the system and some reports are available remotely.

32 Life-cycle Activities

Without input from the users, workload cannot be projected. Even if workload is projected, measurements of capacity remaining are required to predict saturation. The saturation prediction is fundamental to the planning of improvements and upgrades. Further, statistics on feature usage and errors are required to determine if the system is performing as specified or if either program modifications or additional training is required.

Finally, user management must be aware of improvements to the computer technology that they are using. If users cannot spare the time to stay current, they must at least sponsor the effort since they control the finances of the remote installation. Central computer shops track the technology and migrate when the technology is sufficiently mature for their environment. Most end-users do not read computer publications or attend computer meetings and hence do not have the inputs to assess changes in technology. Yet the users should be informed because the continued use of obsolete systems will ultimately affect their ability to compete.

The remaining chapters of this manual address the five development phases discussed in Chapter 1. As each phase is discussed, the design hints appropriate to the task/activity combinations related to that phase are detailed.

3
DEFINITION PHASE

Figure 3.1 combines the task phasing and the activity list for the Definition phase. The principal active tasks during this phase are Project Management, with emphasis on project planning; the determination of Requirements; the Feasibility Study; and the beginning of the System Analysis.

All the design hints given in this chapter may not apply in every case. But even a hint that does not apply may suggest a previously unrecognized area for study or investigation. It is suggested that each reader have a note pad handy as items unique to his environment may occur to him while reading the hints provided. If all personnel at an installation pool their notes, they will have augmented this handbook with items specific to their environment and class of work. Thus designers that follow will benefit from those who have gone before.

34 Definition Phase

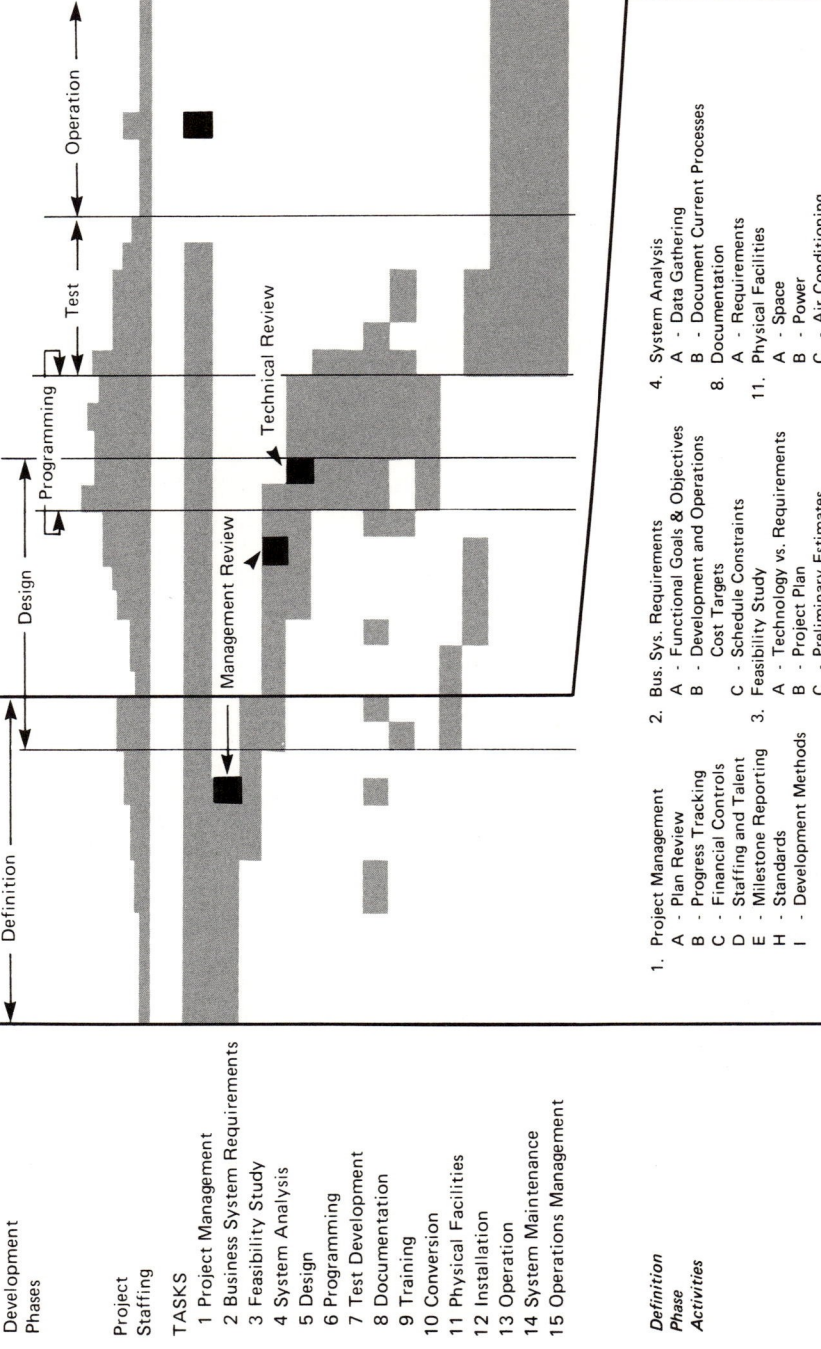

FIGURE 3.1 **Definition Phase**

PROJECT MANAGEMENT

1. Task Chart

As soon as the work requested is assigned to a project leader, he should enumerate the tasks to be undertaken and prepare an initial task chart similar to Figure 1.1.

2. Task Enumeration

For simple projects, list the activities necessary to accomplish each task on a worksheet and annotate each activity with the skill level(s) required, the total effort for the task, and the expected elapsed time. Prepare these estimates neatly as they will need to be revised several times during the project. If they are properly prepared the first time, administrative personnel can provide revisions.

3. Dependency Chart

For large projects, complex projects, projects which are highly visible, or projects to be accomplished on a tight schedule, draw a dependency chart as follows:

A. List and name all tasks and activities.

B. Assign each activity a five digit ID number with the first two digits representing the task number and the last three digits representing the activity sequence number.

C. For each activity, note its prerequisite activities.

D. For each prerequisite, note its dependents.

E. For each activity, note the skills required, total manpower required, the level of manning, machine time, supplies, and equipment.

F. Adjust the sequence and the schedule until the resources required equal the resources available. Increase the resource availability as required if meeting a rigid schedule.

G. Draw a dependency chart (simplified PERT chart) and use it as a briefing aid for management, current employees, and new hires.

H. Draw a summary chart similar to Figure 1.2 for management.

I. Distill the chart and budget for people, machines, and supplies. Use a worksheet similar to Appendix A.

J. Set up a control system to track progress and expenses and to trigger management action to review and revise the plan whenever milestones are not met or the budget appears to be consistently overrun.

4. Peer Review

Once an initial project plan has been produced, have it reviewed by a peer before it is presented to management. In reviewing an early project plan, the peer should concentrate on missing tasks and note any activity estimates which are grossly incorrect.

5. Project Assumptions

In preparation for a peer review, write down the assumptions on which the plan was based so the plan may be revised if any of the assumptions are violated. Specifically, the assumptions may concern the skill levels for the project personnel; the stability of the hardware and software to be used for development; the maturity of the project team as a working unit (i.e., recently assembled from available individuals versus a mature team who have worked together previously); and the amount of interference and distraction expected from other work not directly related to this project.

6. Outside Resources

When planning a project, identify each activity requiring resources not totally under the control of the project leader.

Thus activities requiring support from a central systems programming department, user assistance in file building, plant engineering assistance in preparing a physical installation, or assistance from vendor installation specialists should be separately identified and managed appropriately.

7. Deliverable Items

After the plan has been assembled, flag those activities that naturally produce specific deliverable items such as reports, estimates, specs, running code, user demonstrations, etc. Whenever one of these milestones has been achieved, record actual expense versus the budget and actual completion dates versus the schedule.

8. Percent Complete Reporting

If some administrative mechanism insists on percent complete reporting, prepare the percentages as requested, but do not trust percent complete estimates except where progress is explicitly measurable as a function of a deliverable item (as described in Item 7).

9. Progress Tracking

If Dependency Charts were produced as described in Item 3, mark progress directly on the chart with a yellow felt tip pen using the following rules: Place some color on every box depicting an activity that has been started. When the activity is approximately 50% complete, color one half of the box. When the activity is complete, color the box completely.

Progress can be tracked by watching the pattern of completely colored yellow boxes crossing the chart. A large project with many parallel paths should have many parallel yellow lines proceeding across the chart. If a series of activities is delayed, their progress line will lag and these activities can be given special attention by management.

10. Financial Controls

Before devising financial controls over a project, consider the problem of estimating new work. It would be useful for each installation to have records on each project's cost by task. Then when a new project was to be estimated, you could review earlier task/activity sequences and if they were similar, go back to the historical accounting information to find out what this sequence cost before, and dig into the project notebooks to review the assumptions which molded the previous project. (See Item 5.) If this kind of financial discipline is useful, then financial controls should be set up to track expense by task. Further, project personnel should be given a list of charge numbers so they can split their time when they are working on multiple tasks.

Note that tracking by project is too gross and does not yield an estimating data base for subsequent work. Also cost tracking by phase is too uncertain due to schedule compressions and expansions which move activities from phase to phase.

11. Cost Tracking

For good project control, cost tracking and progress tracking must be coupled. Many installations have good cost accounting systems, but have not set up progress reporting systems of equal sophistication. Tracking expense without tracking progress results in a bureaucratic pursuit of budget status without providing insight into the status of the work.

12. Optional Plans

For very large projects involving heavy expenditures, a project leader should consider gaining access to an automated PERT program for project tracking. If the project is large enough to warrant automated tracking, then the project manager should probably initially prepare three versions of the project plan for discussion with management.

The first version of the plan should assume infinite resources, as this will minimize the schedule time and provide the most parallel paths during each phase of the development. The second version should attempt to minimize the development expense. In preparing this version a project manager will need to avoid peaks and valleys in staffing and be sure that the total resources required in any category do not exceed the resource availability for that category. All of this will result in a longer schedule. The third version of the project plan should be the project manager's best professional compromise between the other two, a trade-off of schedule time and development expense in some manner that fits the project situation.

When the plans are presented to management for review, the first two plans will establish the practical extremes and set the stage for an intelligent discussion of the trade-offs between resource availability, cost, and time.

13. Project Staffing

When staffing a project, be sure the talent levels called out in the project plan are met, or go back and replan the project based on available personnel. If some critical task is at the fringes of the art and requires top talent, do not try to make do with more personnel having lesser skills.

14. Critical Skills

If certain critical skills such as communications, data base design, or system tuning are not readily available when required, set aside some time for education or training or acquire these skills temporarily from outside sources.

15. Personnel Evaluation

After personnel are assigned to the project, consciously evaluate the performance of each person to see if he can produce in accordance with his skill level. Revise the project schedule and/or restaff the project if the talent available does not match that which was assumed when the project plan was assembled.

16. Milestone Reporting

Identify those activities that produce major deliverable items and select a series of them as project milestones. Financial systems usually collect costs by calendar month. Projects seldom have noteworthy accomplishments by calendar month. Avoid mechanical monthly reporting and instead substitute milestone reporting where product completion and progress are known and can be easily measured and demonstrated. Identify these milestones in your initial management briefing and plan subsequent management briefings around them.

17. Project Status

Use the project Dependency Chart that has progress marked upon it (See Item 9.) to open each milestone meeting; thus, busy senior managers can quickly reappraise themselves of the project, its tasks, and its activities before status is discussed.

18. Plan for Change

Planning for change is more critical in a distributed environment than in a batch environment. There are finite physical limits on the number of messages that can be transmitted over a communication line and the number of transactions that can be entered from a single terminal. If the business unit being supported fluctuates in size, reconfiguration of the computing system may be needed to accommodate growth or to reduce cost. An on-line system is much more closely coupled to the fluctuations of the business unit than is a batch system. If the business unit is split, then the computing system may need to be extended or replicated. If two systems are supporting a manufacturing operation and additional manufacturing facilities are built, terminal placement or communication lines may need adjustment even if the total volume remains the same. Thus computing system planning must be kept in step with the planning of the business unit. A plan for change should address the following items:

A. Programming standards to provide table driven code, good documentation, and good diagnostics.
B. Utility programs to create, display, and modify control tables.
C. The ability to temporarily immobilize the system, a node, or one or more terminals and then change configuration tables or edit rules.
D. Traffic measurement, service level measurements, and trouble measurements recorded in machine readable form together with an analysis program and some procedures allows the gathered data to be massaged so problems can be predicted and/or recognized.
E. Modular hardware, programs, data files, documentation, training, and controls arranged so that reconfiguration is relatively easy and the resulting system is still efficient.
F. Flexible computer hardware, communications lines, power, air-conditioning, and site layouts arranged so the hardware and communications subsystems can be reconfigured as required.
G. Plans for an administrative change control system covering:
   1. Standards for design, programming, and testing.
   2. Production application libraries.
   3. Data base definitions and structures.
   4. Information exchange for node-unique problems.
   5. A change control procedure providing information on and controlling the introduction of changes.

19. Project Standards

Any project that is going to support a business unit over an extensive period of time needs standards. The following items are categories of standards which should be considered:

A. Programming language (COBOL, FORTRAN, BASIC, RPG, etc.).

B. Operating system interface.

C. Security classifications and authorization rules.

D. Job and data set naming conventions, version numbers, and change control.
E. Communication message formats (protocols and link disciplines).
F. Data structures.
G. Data element definitions.
H. Data formats.
I. Coding standards (must support test and diagnosis).
J. Interfaces between program modules.
K. Internal program structure.
L. Documentation series.
M. User message syntax, abbreviations, and formats.
N. Test tools and compatible application programming.
O. Priorities for production processing, test, and emergency action.
P. A series of cost accounts for machine accounting purposes.

20. Administrative Controls

The initial development schedule will be based on estimating coefficients which were dictated by the initial set of project assumptions. If technical changes to the assumed hardware, software, or procedures are made, and if these changes affect the coefficients, administrative control procedures should alert project management so the schedule can be revised as required.

21. Design Classifications

As the application is being defined, a preliminary design will start to emerge. Each feature will probably fall into one of the following four classes: software extensions, communication functions, data base management functions, or applications specific functions. Keeping these functions separate simplifies the design and clarifies thinking.

22. Report Mock-ups

When working with users who are not familiar with a computer, good practice dictates that mock-ups of reports and terminal displays be prepared so the user can picture the output products from more than just narrative descriptions.

23. Preliminary User's Manual

Another good practice suggests that a draft of the user's manual be prepared after the preliminary design and prior to the detailed design. Preparation of the user's manual on this early schedule tends to make designers more sensitive to man-machine factors and provides a user-oriented product for his review and concurrence.

24. Test Considerations

Since testing, in one of its varieties, usually amounts to about 50% of the development cost, testing must be given early consideration. Consider the following:

A. Driver programs for unit test that are table driven rather than custom coded for each test.
B. Dump programs that are driven by a dictionary so formatted dumps are readily available.
C. Statistical tools that report on the programming paths tested in a given run.
D. For systems involving extensive communications or high transaction volumes, a test tool that stores results on disk and then compares a new set of results with the previous set of results to detect changes.
E. Good systems require regression test tools:
    1. A library of test cases.
    2. A library of test procedures.
    3. A library into which test output is placed.
    4. Data reduction programs to digest the test output and highlight significant events.

25. Modular Structure

If the front ends of transaction-oriented systems are designed with separate modules for data capture, data edit, and transaction scheduling, then the data capture module (the module most likely to contain hardware dependencies) can be replaced with input driver modules to simulate keyboards.

Similarly, calls to terminal control software can be replaced with calls to a data logger so the outputs can be recorded rather than displayed. These records can then be analyzed by the data reduction program.

Design Note: When the data reduction program is built, it must have flexibility to accept output displays in a different sequence than that in which the related input transactions were presented to the system since internal queueing delays may prevent input and output sequences from being in one-to-one correspondence.

26. Log Outline

Design the data and the transaction log file with testing in mind. If the time of last update is explicitly stored with the record in the data base, then a data reduction program can be invoked following a test which selects out of the data base only those records that were changed. Alternatively, put the fully qualified file key and a transaction type code on the log tape so it can be processed to find the file keys for those transactions which updated the data base. This string of keys can then be used to select the changed records from the data base.

BUSINESS SYSTEM REQUIREMENTS

27. Goals of Distribution

As soon as a systems designer has finished his initial indoctrination and the application begins to take shape in his mind, he should crisply enumerate the goals and objectives he is trying to achieve by distributing the application programs and files.

Once those goals are clear, the designer should take a few hours to define how success will be measured (productivity, elimination of out-of-stock-incidents, reduction of inventory on slow items, control of bad debts, reduction of delay, etc.). Once the measures of success are described, the system designer can ensure that the appropriate times, event counts, and data values are captured during the normal course of transaction processing.

In designing an improved system to replace an existing system, be sure to collect transaction statistics; response time measures; current costs; storage volumes; statistics on errors; and statistics on system failures, blackout time, and elapsed time until back in operation. Since users tend to compare a new system with the previous version, be extremely cautious about pursuing cost savings if the service level goes down in any category.

28. Interviewing Users

System designers must exercise care when interviewing users to determine system goals and objectives. If the system goes much beyond simple data capture or status inquiry, careful interviewing techniques will disclose that different goals and objectives are held by different levels of management within the user's organization. Frequently junior managers will emphasize some particularly painful process in the present system and exaggerate its importance. If one continually asks questions about the frequency of occurrence and relates this information to the total of all transactions, one can avoid biasing the system design to the concerns of one eloquent member of the user's staff.

A different bias creeps in when interviewing senior members of the user's staff. Junior personnel usually live in the present and are uniquely concerned with today's problems. Senior members of the staff live in the future and hence are a good source of information about where the business unit may be going and how rapidly it may get there. However, a designer should not be misled by their seeming lack of concern with today's problems.

Occasionally a computer system fails often enough or hard enough for the users' complaints to reach senior management. Following one of these episodes, the senior manager is very sensitive to computer system failures and tends to emphasize these problems when interviewed by a systems designer. The designer must listen patiently and then use his own professional judgment to place these complaints in the proper perspective. To establish this perspective it helps to ask how often these problems have occurred and the dates on which the problems were brought to the attention of the senior manager. As in many endeavors, managers tend to remember painful experiences long after the problem has been solved.

29. Analysis Outline

The development of business system requirements is an art form. Many analysts have never participated in the processes that take place prior to the preparation of a preliminary design and programming specs. However, with distributed systems, projects can be smaller, and small projects require the project leader to be a jack-of-all-trades. Specifically he may be required to start with a vague work request, develop his own requirements, prepare the justification, pare the project so the portion scheduled for automation is justified, and then proceed through the familiar design, development, installation, and operation sequence.

Until costs and time constraints are introduced, data processing projects tend to include a mixture of mandatory functions, desirable functions, wishes, and dreams. The seasoned systems analyst will make a list of all personnel likely to be users of the system to be developed. The system's sponsor should be able to identify a thoughtful, well-balanced individual in each of the using organizations who can be interviewed as part of the requirements determination process. Sometimes the sponsor cannot identify the exact person, but will know someone who probably knows the person most qualified to speak for a group of users and their interests.

After specifying those persons to be contacted and preparing an agenda for each interview, the systems analyst is ready to start. During each interview he should collect any available reports, memos, or specifications. By reading this background material as it is received, the interview agendas will improve and the system requirements will start to materialize. Some subjects will require in-depth treatment and a list of individuals with the appropriate interests, skills, and qualifications should be built as part of the interview process.

About mid-way through the interviews, the designer will be able to enumerate the primary inputs and outputs, the main automated files, and sketch the principal processing to be performed. In addition, requirements for file building or conversion and the potential impact of new system designs on the existing organization will start to appear. If the analyst writes up each interview and sends it back to the person interviewed for approval, he can guard against being misled by what he thought he heard.

30. Preliminary Design

Shortly after the first set of interviews is concluded and a sketchy preliminary design begins to appear, the analyst will be able to identify patterns of activity which were incompletely specified by the interviews, but which should be exhaustively treated. Thus the design will expand. In addition, the analyst will have a feel for the user's economic situation, the basic development schedule that would be acceptable, and can start to make value judgments between dreams, wishes, desires, and necessities. It is the analyst's obligation to make these preliminary categorizations.

If there is an existing system, the functions it supports tend to be considered necessities. Unless there is a major benefit from deleting some of these functions, they probably should be considered mandatory since the user's organization is structured around the functions now available. The user will also treat the deficiencies in the present system as necessities. The designer must exercise his judgment to determine whether this is in fact correct. Depending on the certainty of future growth and expansion, functions to support growth may also be necessary.

31. Cost Tradeoffs

After the designer has listed all the necessary functions of the system, a preliminary estimate of development and operational costs can be produced. This will be the system's base cost. The desires and wishes will usually sort themselves out when the designer starts to assess the incremental costs.

While the incremental operational costs of an additional display screen may not be significant, additional screens still need to be specified, coded, tested, and documented; personnel need to be trained and in some cases additional data elements need to be captured, edited, stored, and indexed to support the display. By placing costs on the optional features and by estimating the impact of those costs on the development schedule, the designer will be able to pare down the user's requests so a reasonable system results.

When performing preliminary cost tradeoffs, extreme caution should be exercised since some system features impact the overall development costs in a nonlinear way. If a stand-alone system were being considered and the user insisted on a no-break operation, then dual systems with uninterrupted power supplies would be required. Similarly, if two peer-connected systems were proposed and if the data base was split between the two systems with no duplication, there would be times when the local system was up but it could not access data in the peer system due to some malfunction. If this is absolutely unacceptable, then the data bases would need to be duplicated with 100% data redundancy, and a more sophisticated control structure would be required to keep the duplicate data bases in step with transactions originating at both ends.

Since the entire development effort will be oriented towards satisfying the final set of requirements, the professional designer should question the user to be sure that expensive requirements are in fact justified; that the person attesting to that justification is aware of the potential costs; and further, that the person asserting that the feature is justified is authorized to commit the firm to the course of action he has chosen.

32. Preliminary Operational Costs

After the requirements are set, the designer can configure a system, amend his hardware and software order as required, revise the development schedule, and prepare a preliminary estimate of the operational costs for the completed system.

Sometimes the unit cost per transaction, per report, and per terminal hour are appropriate design calculations. They can then be multiplied by the preliminary transaction volumes to get estimated billings. Thus the user could know how much of his operational costs were fixed, how much were variable, and how the variable costs changed with volume.

33. Scheduling Constraints

Schedule constraints imposed by the user are likely to be the subject of difficult negotiations. Much of the popular literature maintains that simple applications on small dedicated computers are quick and easy to install. While that is true, there are no universally accepted definitions for simple, small, quick, and easy. Even if there were, the user may not be sufficiently knowledgeable to be familiar with the standard definitions.

Most often inexperienced users do not understand the development process and frequently have cost and schedule targets in mind months before the requirements are written down. If the combination of hardware, software, and design skills allows a user's schedule to be easily met, then the discussion is over. However, if the job is bigger than the user anticipated, or if the resources cannot be made available to accomplish the work on the desired schedule, or if the programming can be completed but it will take two additional months to build the files, then negotiations are in order.

Sometimes both sides can give and a compromise schedule is readily achieved. Sometimes additional funds are available and additional resources for development can be acquired. Sometimes the system can be developed in increments and critical functions can be installed on an early schedule. Sometimes the user has processing peaks due to model changeover, new product introduction, or annual peaks in his manufacturing cycle. When this occurs, if the system cannot be installed so the shakedown period is completed before the annual peak, then it is frequently better to slip the schedule and install a more thoroughly tested system after the peak has passed.

FEASIBILITY STUDY

34. Physical Limitations

Every computer system has limits. After the requirements are known, the designer should check that the total number of terminals required can be supported by the system under consideration and that the geography and physical placement of the terminals does not exceed cable length limitations.

35. Disk Capacity Estimates

The designer should then make gross estimates of record size for each record type and the number of records of each type. If records must be accessed on two or more keys, estimates for secondary indexes should also be prepared. If these early gross estimates exceed 50% of the disk capacity on order, a precise estimating effort should be initiated, because in addition to these primary files, the system must have space for buffers, queues, logs, programs, software, and working storage.

36. Communications Line Loading

As soon as the number of messages to be transmitted around a loop or over a communication line can be estimated and the length of the average message guessed, communications line loading should be checked. If more than 20% of the bandwidth is used by these early estimates, queues are likely to result and message priorities may be required. More than 50% usage implies an extremely heavy loading that may be incapable of properly supporting man-machine dialog.

37. Application Feasibility

If the application is unique to one locality, if all the inputs and outputs are generated and used locally, if the system has low transaction volumes, and if the system serves only one portion of the business enterprise, it is likely to be easy to install provided the hardware configuration has sufficient capacity for the load. If any of these constraints are violated, additional work will be required to determine the application's feasibility.

38. Nonstandard I/O

Every systems vendor supports a complement of I/O devices in his standard product offering. Further, additional I/O adaptors can usually be ordered so that plotters, custom terminals, and real-time devices may be connected. If nonstandard I/O is required to support the application, feasibility cannot be determined until the hardware interface and the data streams across that interface are determined for each device and reviewed by a vendor specialist.

In addition to hardware attachments for special devices, some software modifications will be required. In most cases these will not be extensive, but they will bring with them additional administrative and support chores so the local software extensions can be prepared and maintained in a manner compatible with the vendor's standard software offerings or programming fixes.

39. Performance Estimating

In some cases the amount of processing required for an application may severely tax the capacity of the installed system. If the designer is faced by heavy computing loads, critical portions of the application should be pilot tested. Sometimes vendors have performance tools which can aid in estimating CPU loading, and critical cases can be programmed and run on real hardware so actual timings are available.

40. Capacity Loading

An application that taxes the installed capacity is always troublesome, but small machines seem to saturate more abruptly than do large host processors. Thus a designer trying to fully load a small machine must design with performance in mind from the very first. Unless the project team has negotiated special support from the vendor's forces, full capacity loading is not recommended until the project team is proficient with the system and its software.

Experience indicates that small computers tend to become disk bound before they run out of cycles. I/O counts, file organization, and the interaction between high speed and disk storage seem to be universal problems when applications tax the capacity of a small machine.

41. Distributed Data

Some applications will require data to be accessed from two or more locations. While two computers may be connected with communication lines and either machine may hold the data centrally or distribute it somehow, it is not always possible to economically implement a designer's first design.

Distributed data is new to the computing community, and while simple distributions are easily accomplished, more complex distributions require careful design and implementation. Later portions of this manual will indicate how complex designs can be constructed. However, during the Definition Phase of a project, data is insufficient to guarantee that any particular design consisting of hardware, software, and distributed applications will run successfully and economically. Caution therefore is urged in the Definition Phase as feasibility cannot be assured until part of the design work has been done.

42. Installing Change

The feasibility study contrasts the available technology with the requirements. Part of that technology is the hardware and software base supplied by the vendor. Another part of that technology is the skills available in the project team. The third part is the operational and management talent available in the user organization. Systems can be built that are too complex for the user to successfully operate. It is also possible to build systems that the user management is ill-prepared to administer. If the requirements imply a drastic change in the way the users conduct their business, the designer must determine whether the users are willing and capable of making that change. If the answer is yes, then the system must be carefully designed, and the development process must provide sufficient documentation and training so the users can assimilate the changes.

If a major change in user operations is dictated by the goals and objectives set by management, then such change must somehow be accommodated. If it can be determined early in the process that an initial design will be too complex or the rate of change will be too great for the users to assimilate, then alternate designs can be prepared. These usually involve more applications programming and sometimes involve more hardware. Barring this solution, the system can be developed and installed incrementally, with the first simplified system installed early and additional increments of function installed only after user personnel have become comfortable with the previous version. This strategy allows major changes to be undertaken by controlling the rate at which change is introduced.

43. User Organization

In addition to tuning the system design to the user's capabilities as described in the previous item, sometimes the user's organization and the support organization also must be tuned so production operations and continuing service can be successfully conducted. When processing and data are distributed, responsibility is distributed. If user management is too busy, unwilling, or incapable of accepting the responsibility for system administration, service, security, training, and some of the responsibility for problem determination and recovery, then a distributed system is likely to be unsuccessful. Education, training, and user participation in the design are the best ways to avoid this problem.

44. Application Portability

Some businesses require extreme processing flexibility. In a distributed system this can be achieved by designing programs so that processing modules and data can be moved from peer to peer, or from host to node in hierarchical configurations. Appendix B contains a sample difference list which shows how to contrast language dialects available on the smaller systems with their host counterparts to determine if enough compatibility exists to support practical levels of portability.

45. Terminal Control Software

If an existing computer application is being redesigned, and if some of the host data and processing will eventually migrate to a remote site, the designers should investigate the use of compatible terminal software. Such software allows the small machine to emulate the functions of remote terminal controllers so a remote system can be initially installed without changing the host applications. Once the remote system is installed and operating, outboard editing and data storage can be enhanced in easy increments until the correct balance between central and decentralized data and processing is achieved.

46. Multiple Terminal Sessions

Several vendors offer network software with features attractive to the designer planning either peer-to-peer or hierarchically connected nodes. Some of this software contains sophisticated protocols which allow multiple simultaneous independent sessions to take place over the same communications line, e.g., split a physical link into several virtual ones. Two or more applications can maintain separate sessions and transmit transactions and responses over a common communications circuit with little interference and no custom communications programming. Additional sessions can be used for operator to operator communication, problem determination, alert messages to the system operator, downline program load, and remote callout of statistics maintained at a remote site. These features aid in maintaining system control in an application design which requires distributing data and processing.

47. Amending the Plan

After the available technology is matched to the applications requirements, the designer may need to amend his original project plan. Specifically:

A. If the application taxes the capacity of the projected equipment configuration, additional technical activities must be added to the plan allowing careful assessment of performance and capacity limits in order to contain risk factors.

B. If nonstandard I/O devices are contemplated, additional training and some outside help may be necessary if the system is to be built on a reasonable schedule.

C. If the changes required in the user and his organization are too great to be assimilated in a single step, and if the system is to be installed incrementally, then the last half of the project plan must be redone. The design must address versions with ascending increments of function and the plan for programming, test, and operation must deliver and install those increments. Furthermore, additional conversion effort may be required to assist the user in migrating smoothly from version to version.

D. If a sophisticated design for distributed data and processing is contemplated, the designer may elect to add effort, seek outside help, and schedule an additional technical review before the design is released for programming. This could change the overlap between the phases and stretch out the project schedule.

48. Amending the Estimates

Regardless of whether the plan changes, the designer may wish to revise his estimates after he has compared the application's requirements to the available technology. Naturally if the plan changes, the estimates must change. But even when the plan holds firm, the designer may note that some activities are closer to the edge of technology than originally contemplated. For activities that require skills which are not common and routine, the project leader may decide to hold to the original plan and schedule, but change the level of talent applied to certain critical activities, or to bring in outside talent possessing experience unavailable in-house. Either of these alternatives will change the expense and payout.

SYSTEM ANALYSIS

49. Collecting Existing Forms

A designer will very seldom be faced with the problem of developing a new system from scratch. Most systems developed today automate manual systems or in some sense enhance existing automated systems. Therefore knowledge about the existing system is available and should be collected, organized, and analyzed as part of the systems analysis process.

As discussed in Item 29, the analyst should collect input forms and reports as users are being interviewed. After collecting the forms and reports that are readily available, the analyst should then build an exhaustive set. If the company has a forms control section, a visit to the control clerk with the forms already located will frequently allow a complete set to be built quickly. After a complete set is built, a revisit to the work space occupied by present users may uncover some forms overlooked or some draft forms not yet submitted to forms control for numbering and reproduction.

Similarly a visit to the production control section within the computer center will provide access to the reports distribution list. This will allow the analyst to quickly fill out the missing reports in a report series and if the list exists in reverse sort, will allow an analyst to determine all the reports received by the user personnel he interviewed. After the report samples are collected and organized, a visit to the work space occupied by present user personnel may disclose hand-prepared reports which contain a blend of manual and automated information.

50. Collecting Reference Lists

On every visit to work spaces occupied by present users, the analyst should look for ready reference lists held on microfiche, in notebooks, on a Rolodex, in a 3x5 card file, or under a glass-topped work surface. Sometimes these lists are posted on the wall above the worker, on the sides of file cabinets, or on pull-out arms of desks. Whenever they may be found, these contain code conversion tables, indexes, or conversion tables to assist in manual processing. A one-to-one correspondence is likely to exist between these tables and tables in the computer program you are about to design.

When such tables and reference aids are discovered, part of the designer's work is already performed for him. One of the designer's activities is to separate coefficients which are likely to change from fixed processing algorithms. Whenever he finds a book of procedures or standard operating practices and a separate table of prices, coefficients, or standard codes, part of that design separation has already been performed.

51. Documenting the Existing System

If an existing system is being replaced, the designer should seek and/or create high level process flowcharts for the existing system. On copies of these charts he can then annotate the volume of transactions flowing through each part of the system, the number of people performing the work, their skill levels and wage rates, and the performance of the present system including time delays, error rates, backlogs, and proportions of transactions being expedited or receiving priority. Given an understanding of this information, it is possible to note where every transaction originates, where every report is used, and the size and contents of all standing files.

Many analysts proceed without thoroughly documenting the existing system. They usually encounter a series of unfortunate surprises since they will have overlooked an important exception procedure or a priority path.

When a new system is delivered, it is human nature to compare the function and performance of the new system with the system it replaced. Unless the old system is documented in moderate detail, new system acceptance may be inhibited since the new system may not provide a necessary function or may offer reduced performance in one or more critical areas.

When the existing system is documented, the applications designer should review its detail with the application sponsor and with supervisors responsible for all portions of the existing process. These reviews will yield multiple benefits: they will verify that the designer understands the existing processes; they will provide the present user population with documentation that can be used to train new employees and brief their own management, while they continue to operate the old system and wait for the new system to be developed; they will provide a vocabulary and nomenclature for the designer to use in discussing system changes; and they will provide the baseline documents so conversion planning can be initiated.

52. Document Special Procedures

In documenting the current system, the designer should be careful to inquire about all routine periodic processes. Sometimes the patterns of flow change when the books of account are closed each month, when physical inventory is taken, when new products are introduced into the production line, or when the production of old products is discontinued. These same events will occur in the new environment, and the new system must contain provisions to accommodate them.

53. Communications Traffic Analysis

When gathering data for a system which has intense communications overtones, a source-sink traffic analysis is mandatory. Prepare a matrix showing the type of transactions that can originate from each node, the average daily volume of those transactions, and the peak hour volume anticipated.

Now take these transaction volumes and add destination statistics. If the flow of information from originating location to destination can change due to time of day, time of month, or some other activity pattern, multiple source-sink matrices will be required.

Once the transaction routings have been determined, go back and describe each transaction by indicating its minimum, average, and maximum lengths.

Analyze these worksheets to find the worst case peak hour for each location. Multiply the transaction mix flowing to or from that location by the proper lengths to find the absolute minimum communications capacity required to handle the load.

Review the above analysis and take note of any predominant patterns. In many cases the daytime flow of information is predominantly from the periphery towards the host. In some cases the system has a balanced dialog during the daytime hours, i.e., inquiry-response. In other cases the host is downline loading data or programs during the nighttime hours. When growth is considered, one frequently finds all three modes of operation exist over time.

DOCUMENTATION

54. Vocabulary Controls

When documenting a plan, a requirements analysis, or a feasibility study, keep in mind the background and skills of the intended audience. Avoid abbreviations, acronyms, and jargon if the documentation is intended for users or managers.

55. Documentation Techniques

Most projects benefit from a controlled vocabulary. Big projects are easier since they usually have a full-time technical editor to process the documents before they are published. The editor will usually maintain a project glossary and ensure that the documents are written according to a consistent vocabulary. Smaller projects must try to achieve these same benefits without the editor's professional help.

A table of contents in the front and an index in the back improve any technical document. Most text processors and some word processors assist in the preparation of these indexes.

Furthermore, a medium-to-large project can justify automated text tools merely to maintain the documents as the project progresses.

56. Document Assumptions

One admonition bears repeating. Always write down your assumptions and where appropriate, place them in the front of project documents. Occasionally one finds truths that are not obvious, assumptions that are in dissonance with the business plan, or facts that cannot be substantiated. When assumptions are stated as part of the documents produced during the Definition Phase, they get reviewed and exception will be taken if any are in error or are controversial.

PHYSICAL FACILITY

57. Space

Sometimes casual decisions made during the Definition Phase return to haunt a project. Although small computers operate in a normal office environment, one should not ignore its space requirements altogether.

A. The space must be adequate for the equipment to be installed with some margin for growth. The equipment cannot be packed so tightly that vendor field personnel cannot service it. Even though the operational configuration may be severely restricted with no additional hardware to spare, it may be desirable to install extra terminals, printers, or other support equipment for development, test, and initial production. This equipment is easier to install and de-install if sufficient space for the augmented configuration was provided in the initial planning.

B. The power requirements of most small machines are nominal, although like all electronic equipment they are sensitive to abrupt power fluctuations and voltage spikes. As a general rule a small computer should be on a circuit which does not contain elevators, punch presses, or other heavy equipment.

If the installation requires extremely high availability and if special power is being run to the computer, you may need similar circuits for emergency lighting and for the terminals supporting the high availability application. Thus the entire computer system and its crew of on-line operators will be available during times of emergency.

C. While most small computers will operate in a normal office environment, determine the extremes of temperature and humidity to be encountered in the space planned for the computer installation before special air conditioning is ruled out completely. Air conditioners and furnaces are sometimes shut down for maintenance or over long weekends to the detriment of computing equipment. Offices adjacent to manufacturing facilities sometimes experience vapor, dust, or particulate problems which exceed the tolerances of the equipment or which require the equipment filters to be serviced more frequently than normal to avoid overheating.

If trouble is sensed concerning the space where a small system is to be installed, the project leader should obtain copies of the vendor's installation manual and check the proposed environment against the requirements stated.

4

DESIGN PHASE

Figure 4.1 combines the task phasing and the activity list for the Design phase. The principal active tasks during this phase are Project Management, Systems Analysis, Design, Programming, and Test Development. Other active tasks are Documentation, Training, Conversion, Physical Facilities, and Installation.

All the design hints given in this chapter may not apply in every case, but even a hint that does not apply may suggest a previously unrecognized area for study or investigation. It is suggested that each reader have a note pad handy as items unique to his specific environment may occur to him while reading the hints provided. If all personnel at an installation pool their notes, they will have augmented this handbook with items specific to their environment and class of work. Thus designers that follow will benefit from those who have gone before.

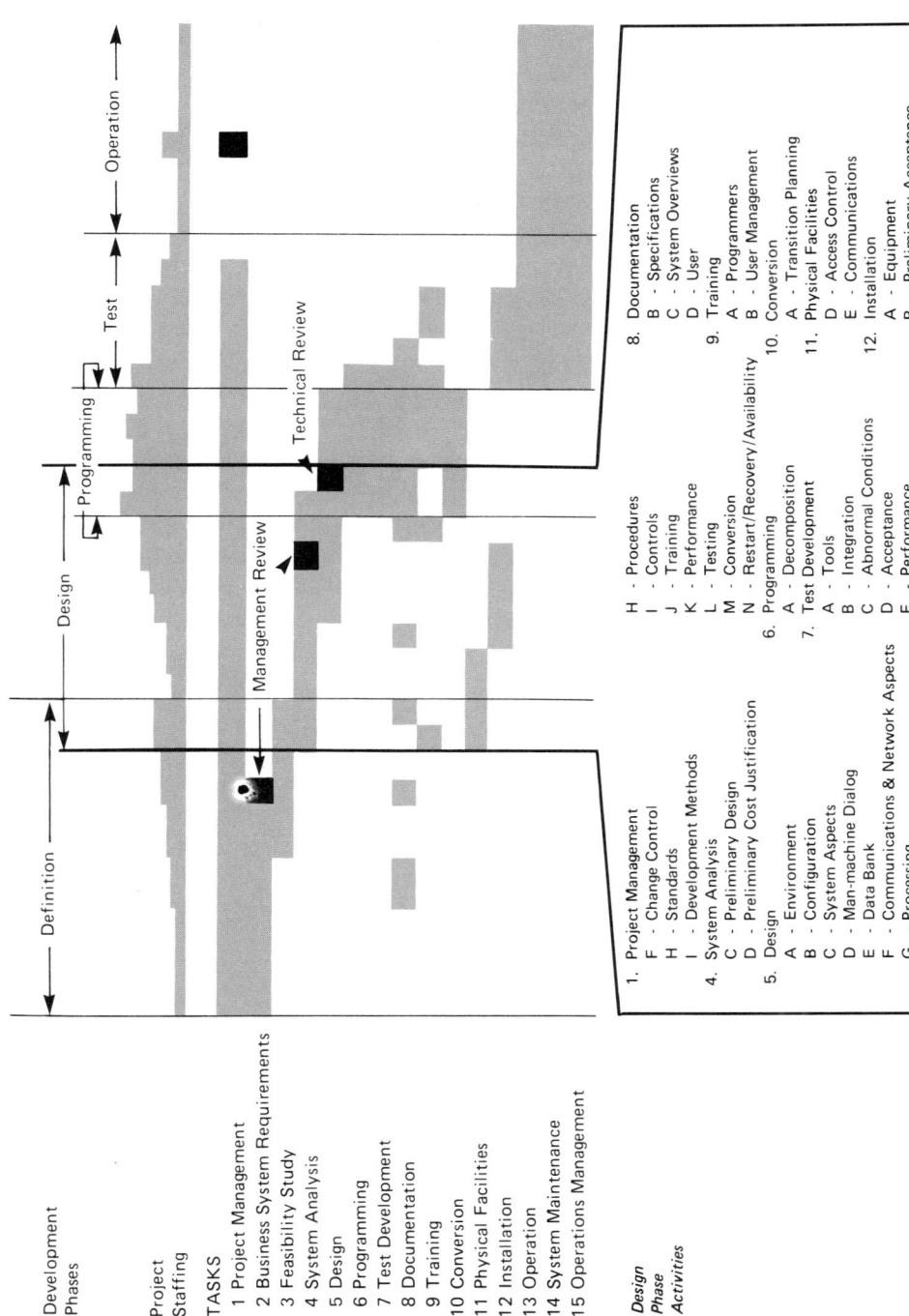

FIGURE 4.1 Design Phase

PROJECT MANAGEMENT

58. Input Conventions

Programming standards should be established so application designers avoid areas that in the past have proved troublesome. For instance:

A. Beware of the use of the following letters for isolated codes as they sometimes cause confusion: I, 1, O, 0, 2, Z.

B. Establish a convention for handling missing data. For instance, double zeros for dollar amounts (00.), a string of dots for alphabetic characters (....), and a dash for an alphabetic field (-).

C. If summary data is presented in ranges, be sure the range boundaries are properly set and treated consistently. For example, $x = o$, 1-9, 10-20 might mean that x can have the values zero; zero is less than x is less than 10; 10 is less than or equal to x is less than or equal to 20; and so on.

59. Name and Address Handling

Names and addresses always prove troublesome. New designers frequently take data entry shortcuts only to find that the style desired for a name and address cannot be obtained without editing and manual intervention for the exceptions, which is almost always more work than if the name and address were entered properly in the first place. Thus some standards for handling names and addresses are in order. For instance:

A. Keep components of name and address in separate fields:
　　(1) Dr. (2) Alan (3) J. (4) Thrush (5) III
　　(6) Central Medical Center (7) Suite ABC
　　(8) 12345 (9) Old Lane
　　(10) Akron (11) Ohio (12) 4xxxx
　Thus 12 fields would be defined.

B. If proper names are to be handled and if these must sort properly for presentation on reports and published lists, maintain two fields: name formatted for printing, and a sort form. (d'Angelo, McDonald, and the A. B. Dick Company appear in the telephone book following Dang, McDole, and in sequence with all the other subscribers named Dick.)

60. Check Digits

Where unique identifying numbers are assigned to credit card holders or vehicle parts, modern technology suggests that each one of these unique identifiers carry a check digit. Further, verifying the check digit *each* time it is handled, in contrast with checking only on entry into the system, is a good discipline. A surprising number of internal processing problems will be revealed by such checking before the erroneous identifier reduces the integrity of the file or interrupts the logical sequence of processing.

61. Control Tables

Good design methodology suggests that all location-sensitive information be eliminated from program code and placed in tables. If the design is so constrained, then the programs are standard throughout the network, the data has a common definition but the values are unique to the local site, and only the tables need to be customized when a processing module is moved from location to location. Naturally such a design would require an update module so the tables can be changed and a print module so the tables can be printed out for verification purposes.

An extension of the control table idea suggests that a single terminal control table be prepared which contains an entry for each terminal containing terminal type, serial number, physical location, line code, and logical address. Along with this static information would be placed dynamic information such as current status, operator ID, security state, line condition, etc. This table would be used for message verification, routing, and problem determination. The table would be established at system startup and modified by each log-on and log-off. The table should be positioned to be accessible from a remote site so a central service location can assess the instantaneous configuration of the machine during remote problem determination activities.

62. Output Checking

When a transaction processing system prepares output messages or reports such as purchase orders or bank drafts, it is usual to build in logic which compares quantity, amount, requesting location, or other key parameters against a table of limits. The table of limits can be set to provide special controls over such events as big ticket items, items with unusual occurrence, or catastrophe-if-wrong items.

63. Clocks and Lockouts

In any network system accurate time-stamping is required for message precedence and problem determination. If multiple intelligent nodes process transactions and if each has a separate time-of-day clock, then provision must be made at system startup, at restart, and at periodic intervals throughout the day to synchronize the system clocks. This can be done by a priority transaction which temporarily quiesces each node in the system, placing it in an immediate access mode to the node containing the master clock. Then the two clocks can either be synchronized or the standard offset determined.

If multiple applications are running on a single processor, and if one application must seize control of the data base so a sequence of commands can be processed without interruption, or if a single application must usurp a communication line for a long message, the prudent designer will cause both such actions to set a time interrupt prior to the beginning of the action sequence. Then the process can be interrupted and control can be voluntarily relinquished based on the time-out parameter entered when the system is initialized. Thus indefinite lockouts are prohibited and the process requiring abnormal access can voluntarily relinquish control so it can be restarted at the point of interruption.

Some applications have routine processes that are active through the day coupled to end-of-day processes which are activated only when the daily transactions are completed. Some designers have erroneously elected to make the end-of-day process time dependent so it wakes up, checks the time-of-day clock to see if the time has arrived for processing, and if not, sets a long cycle interrupt and goes back to sleep. Such a design is not fail-safe since any failure that interrupts the daily processing may delay the processing of the last daily transaction. If this delay is sufficient, the summary process can wake up and start to execute while there are transactions still in the queue. One fail-safe approach would be to cause the transaction processing application to place a specific wake-up command in the processing queue after it processed the last transaction in the daily stream. Thus the sequence would be assured even if the daily processing were delayed.

64. Coding Standards

Other programming standards are frequently set to control the way code is constructed and data is handled. For instance:

A. Always write deleted records on a log file along with time, date, ID of the process causing the delete, and, if several individuals are authorized to use the process, the ID of the individual deleting the record.

B. Since failures will occur, it is best to make provision for them. Thus when processing transactions, many designers retain the message ID in a standard location within the data area. Then any module that finds it necessary to write an error message on the log tape can go to a single place to pick up the ID of the message being processed.

C. Other designers set programming standards so working space is used in an orderly progression and then overlaid starting with the oldest fields first. Then if the process crashes, a snapshot of this area will aid in problem determination.

65. Reports Standards

A set of standards can be established to provide uniform-looking reports so the population of administrators and clerks see a comforting consistency regardless of who designed the specific application. For instance:

A. Good report design dictates that the title, unique ID number for the report, the date the report was printed, and the page number should be on each sheet of every report. Furthermore, the last page of the report should carry meaningful item counts and control totals as a notation that the report is complete.

B. If the components of the total are printed, then print the total.

C. When printing lists that contain control breaks, provide extra space following the break and as a general rule provide both item counts and dollar totals. Be sure to provide grand totals at the end of all complex reports.

D. Reports should be designed with consistent (standard) units of measure and consistent scaling. If standards have not been established, or if the units and scaling must vary, then the units and scaling must be printed with each value.

E. When printing decimal amounts, line up the decimal points rather than left justifying or right justifying decimal numbers.

F. Every report should be in some order. If manual files or processes are involved, the report should be sorted to match the order of the manual process. For example, if manual pending files are in last name alphabetic sequence, then the computer reports should be sorted to last name alphabetic sequence.

G. The performance of some printing devices deteriorates if the special symbols in the character set are used heavily. The designer must be aware of the performance characteristics of his output devices before the field encodings for coded variables are established.

66. Dialog Standards

If CRT terminals are being used by clerical personnel for a high volume activity, then human factors standards are in order. For instance:

A. When using CRT screens for monitoring input data, or for correcting a record, or simply for displaying a record, keep the formats consistent to ease operator training and the skills required. Also make sure that button sequences are consistent even if slightly more keystrokes are then required; the consistency will more than make up for the extra keystrokes.

B. Be consistent in handling options on CRT input screens; that is, phrase questions so that a "yes" answer always implies the same action; provide consistent methods for backing up a record or retrieving the previous screen, for skipping a field, for skipping to the end of a screen, for recalling a previous menu, etc. Set up these basic protocols first and then make sure that each application involving screens conforms to them.

C. Consistency of screen design is most easily guaranteed if standards are established which cover prompting, data entry, data change, data display, data delete, message delimiters, font selection for interactive dialogs, and the like. If the variety of problems is recognized before any screens are formatted, then screens that logically perform similar functions can be similarly structured to the benefit of designers, programmers, and terminal operators.

67. Message Standards

In a communications-oriented system, the standards set for handling message traffic must encompass function, performance, and problem determination. For instance:

A. If messages of different urgency are to be mixed in the traffic across a single communications line, then the prudent designer should place a priority code in the text of each message and maintain processing queues by priority. Thus if the system becomes overloaded, the scheduling of message processes can be changed from FIFO to priority.

B. Furthermore, designers of complex networks have found it necessary to put a message type/class code indicator next to the priority flag in each message. Thus inquiry, response, data for delayed update, program blocks being down-line loaded, system commands for immediate action, etc., can all be differentiated and processed properly even though they may be in the same priority class.

C. Other designers have proceeded one step further and added a flag to each message to distinguish between messages intended for test, training, and production. Thus message descriptors are stored with the text of the message and these allow the queue processing/scheduling algorithms to route messages correctly for processing.

D. Developers of large military communications systems suggest that each node should be capable of logging all message traffic on a storage device or optionally transmitting all message traffic to an upstream node. This logging capability would be used very selectively and when used, each message would be clearly identified as a duplicate. However, in case of problems, the input to the message queuing and routing module should be logged so troubles with queues or priority processing can be diagnosed.

68. Debug Standards

Testing in a network is much more complex than testing in a batch environment. The times of arrival of the inputs and the departure of the outputs are asynchronous; the input and output queues are volatile and have no long term memory, unless these factors are specifically programmed. Further, the configuration of equipment justified for routine operation is sometimes smaller than would be desired for efficient testing and debugging. Thus good program design methodologies suggest that testing be considered before the preliminary design is frozen. For example:

A. When designing a distributed application, design standard debugging tools first, then set programming standards so audit trails, residue, and structured storage are compatible with the tool design.

B. Since multiple applications on small computers sometimes interfere with each other, causing performance to suffer, the designer needs to allocate the hardware resources to the several applications which are expected to process simultaneously. Then to test that each application uses only the resources it was allocated, the designer must specify a general purpose parameter controlled program which has no purpose but to use resources in accordance with a set schedule. If this program is running during the performance testing of each application, it will keep one application from exploiting more resources than it was allocated. Then the other applications will not suffer degraded performance.

C. When test tooling is deferred until testing has commenced, the tooling tends to be ad hoc, error prone, undocumented, and one shot. Money, time, and effort would be saved if the test tooling were designed along with the application; if the tools were table driven; and if the output programs were dictionary driven. Such generalized tools could be independently tested to ensure they were of high quality and would be worthy of documentation so they could be retained for use throughout the life of the system.

D. Some quality assurance software is beginning to appear which inserts temporary counters on each leg of each decision prior to the source program being compiled. If these counters are printed out after each test (or group of tests) is run, then the programmer will know what paths through the program were tested by each test case. Moreover, if the set of test results is reviewed by the project leader, he could determine the degree of coverage of the entire set of test cases and could determine if additional cases were necessary to test all program paths at least once.

69. Central Service Standards

In a large network, one node is frequently assigned the responsibility for network operation. If system availability is critical, special programmers, called Coverage Programmers, are assigned to staff the central network support node. Coverage programmers should be assigned to this node for all shifts with operations critical to the business, and a full set of documentation should be assembled. Coverage programmers would then be available to answer questions, perform preliminary problem determination, work with vendor field personnel if service or maintenance were required, and then to restart and restore normal operations after repairs were made. If such a support center were planned, then additional standards are required to define the relationship between the nodes and the support center. For instance:

A. The nodes must always be conditioned to accept a command for immediate execution from the support center.

B. Every time a node recognized an error and successfully retried the process, history information should be transmitted to a logging program within the node. The logging program could count the number of retries until a threshold was exceeded and then send a service message to the support center.

C. If a node logged errors in a local error table and transmitted on request or wherever a threshold was exceeded, then it must be possible for the support center to transmit an inquiry to the node and get the current contents of the table as a reply.

70. Availability Standards

High availability results from systems with built-in error checking. Hardware designers have long built equipment which contained some checking circuits. Some vendors build in diagnostics which are automatically executed when a system is started or trouble is suspected. Some vendors even provide software systems diagnostics which are automatically called at the time of failure. This software then surveys the status of the machine and upon finding control blocks in error, undertakes to reconstruct the erroneous control blocks so a warm start is possible, avoiding the necessity of a cold start. These same philosophies can (should) be carried over into the design of high availability applications.

A previous section (Item 60) discussed check digits on primary file keys. If control tables are structured, checksums can be placed on them to determine whether the table is internally compatible with itself and can be used for restart. The object code produced by some vendors' compilers is reentrant, so it can be refreshed from disk on a restart or it too can be checksummed to determine its integrity. If input forms are designed with redundancy, item counts are taken with all batches, and special flags are used to indicate the ends of queues and reports, then the designer has made an attempt toward structuring his application for efficient use. Applications diagnostics can then be written to prove the integrity of the application so that the system can be warm started after a failure, and the operators need not undergo the delays and redundant effort usually associated with a cold start.

SYSTEM ANALYSIS

71. Transaction Frequency Analysis

When designing transaction processing systems, most experienced designers start with a transaction frequency analysis. All the data on the existing system is analyzed and all the individual transactions are given a type code. While transactions may be grouped later in the analysis, initially every unique combination of data fields is given a unique typing code. Thus even if the data fields were the same but the priority differed, there would be two type codes.

When modernizing an existing system, actual counts of each transaction by type can usually be obtained. Sometimes these counts were designed into the existing system and are available as traffic analysis reports. At other times they can be obtained from an analysis of billing data or raw usage accounting records. Sometimes the transactions are created on prenumbered forms, and users of forms can be asked to record the number of the next unused form each morning so daily transaction data is at least available. In the worst case a data gathering sheet must be provided for every person originating a transaction, so a week's worth of data can be obtained.

If transaction counts are actually obtained, either by sampling or self-recording, the analyst must be careful to first validate that the daily transaction volume was not affected by the data gathering process, and second to determine how the data gathered relates to the normal data flow.

Some work units process from a backlog of data and hence the daily transactions will tend to vary only with employee morale, i.e., high on Tuesdays, Wednesdays and Thursdays; and somewhat lower on Mondays, Fridays, and the days before and after a holiday.

Other business units will be demand driven, and the transaction volumes on Mondays and the days after holidays are likely to be much greater. Credit and finance industries tend to be biased by paydays and ends of the calendar month.

Some retail enterprises process a volume between Thanksgiving and Christmas which is 25% above their annual average daily volume. But even if the systems analysis portion of a project happens to take place in the spring, data on transaction volumes is still valuable and worth gathering. However, the designer must find a way to relate the data gathered during the off-peak months to the times of peak activity.

When dealing with a system which has 24-hour availability, transaction data is usually gathered on an hourly basis. Not only does the volume of transactions vary with time of day, but frequently the mix of transactions (relative frequencies) also varies with time of day. Airline reservations, law enforcement, and hospital activities all experience a very pronounced diurnal activity pattern.

Once the number of occurrences of each transaction throughout a measurement period is known, transaction analysis should start. First the designer should determine if all the possible transactions are enumerated or if some noteworthy transactions are missing because they did not occur during the measurement period. For example, if the system did not fail during that period, no restart transactions would be counted; if the system did not fail hard, no cold start transactions (reload the system, determine the integrity of the files, rebuild the files as necessary) would have been counted. It is quite possible that the end of the month close, the end of quarter tax report, or the end of the year activities, including the close, the reinitialization of fields used for cumulative business totals, changing of dates, and the restructuring of the financial chart of accounts may not have been counted since they did not occur during the measurement period.

Thus after the data is gathered, the analyst must be sure that the transaction enumeration is exhaustive. No frequently occurring transaction will have been missed; therefore the data gathered is probably sufficient for performance and queuing studies. However, the infrequently occurring transactions are usually responsible for special system functions and require additional code, testing, and documentation which significantly add to the development effort. Further, they may be extremely long running and even if occuring only infrequently, they impact routine performance.

After all of the transactions have been exhaustively enumerated, and the designer has convinced himself that the volumes are valid and typify the real world, changes in the business environment must be hypothesized and reflected in the transaction volumes and mix.

If a business unit projects growth or contraction, and if it can be ascertained that the customer base will remain stable, then the frequency histogram can be scaled by simply multiplying each transaction count by the growth rate. However, such is usually not the case. When a business grows, new customers are obtained and the frequency of records created increases. Thus the ratio of creates to updates would change. If a business grows sufficiently and new employees are required to handle the volume, the system will see the compound effects of business growth and unseasoned employees. Not only will the number of record creates be up, but the number of error-related transactions will increase relative to the business volume. The error experience of a new branch office can be expected to be considerably higher than the errors experienced by an existing branch office even though the net business volume may be the same.

Furthermore, senior management may plan to change the direction of the business itself. If new product lines were contemplated, a flurry of activity in the manufacturing data base would be expected. Abnormal activity would be expected in all data files related to quality assurance as the product was initially offered and until production became routine. New sales programs and new marketing efforts could be expected to add customers and expand territories, and these in turn would mean additional salesmen, different patterns of sales activity and expense, etc. Thus in designing a new system, a designer must capture today's transaction frequency data, be sure that all transaction types are exhaustively identified, adjust the data based on any routine business growth expected, and readjust the data based on any new product introductions or strategic changes in direction foreseen by senior management.

However, two words of caution are in order. First, the frequencies of transactions and their mix will be quite different while the new system is being installed and the work is migrating to it. It is hoped that no new function is required to accommodate this migration (other than bulk file conversion), and that the volumes of additional transactions due to initial system instability and staff unfamiliarity will be less than the capacity installed to accommodate growth. Thus the installation transient will use some of the residual system capacity, but after the system settles down, the capacity will be once more available to accommodate growth.

Given a range of expected activity, some designers have made the mistake of adding the largest numbers to come up with a conservative estimate of capacity required. One designer got sales projections from the marketing department which indicated their business volume was expected to double within two years and double again within the next two years. Based on these optimistic projections, he concluded a large computer was required and the economies of operation indicated that it should be purchased. He compounded his error by installing it in a facility that was buried inside a manufacturing plant, inaccessible to nonemployees. When the business did not materialize, the company had too much equipment without being able to share it. Thus expense rose without an offset in revenue or usage. A prudent systems designer, faced by business forecasts which were greater by powers of two, would have designed a system which could have been reconfigured as the business grew without overwhelming expense in the interim.

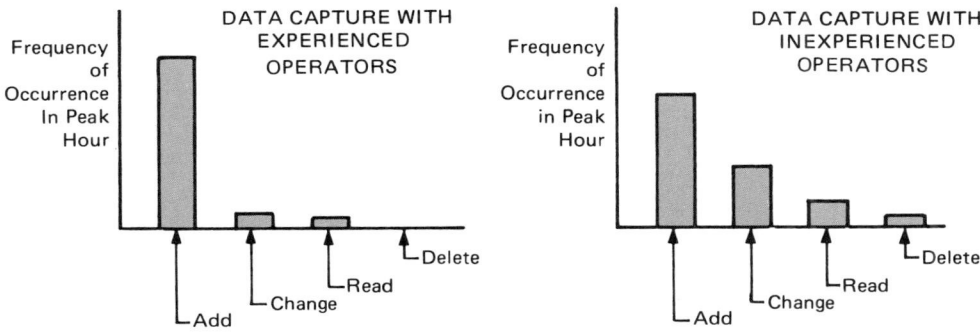

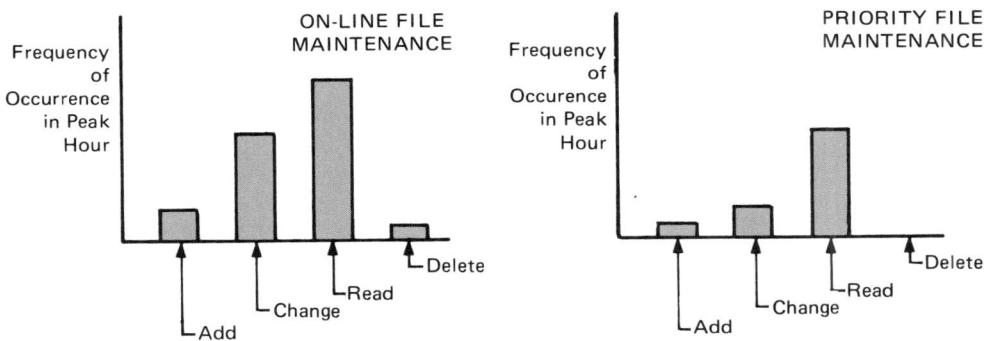

FIGURE 4.2 **Transaction Frequency Histograms**

Figure 4.2 shows some simplified transaction frequency histograms. Rather than depicting transactions as received from a network, they show transactions between the processor and the file system. Thus the four basic file activities — Add, Change, Read, and Delete — are depicted. Note one histogram depicts priority transactions separately from the routine flow.

Priorities always cost resources and capacity. With normal transactions a designer might process all transactions of a given type that reside in the queue (mini-batch), as some host data management systems do, before obtaining the object code to process the transactions of a different type. If a transaction invokes high priority, then it moves to the top of the queue and is processed as a batch of one. Thus all overhead for a minibatch would be experienced just to process a single transaction.

Designers should be cautious of priorities. Any extensive use of priorities (dictated by the applications environment) will cause additional overhead and reduce the average throughput of the system. Therefore priority transactions should be counted and analyzed separately.

72. Transaction Sequence Analysis

After the transaction types have been enumerated and the frequency of occurrence determined, a transaction sequence histogram should be prepared.

The transactions in many systems occur singly, in no discernible pattern. However, in a large number of systems transactions occur in clusters, and while the system must be programmed to accommodate random transaction sequences, a design which favors frequently occurring sequences will perform much better. Thus if one were designing for random single transaction processing, each changed data record would be posted to the output queue at the completion of each transaction.

However, if it were highly probable that transactions occurred in sequences, the record buffers would be checked at the beginning of each transaction to see if the previous transaction had already obtained the desired record from the file.

Similarly, if an order entry system processed a transaction which opened an account for a new customer, the module which allocated storage, assigned customer numbers, and edited name and address could call for and preload the processing module which accepted orders for goods. This can be done because new customers probably would be added to the file as a prelude to the order process. Similarly, in airplane reservations, the module that checks the status of a flight could call for the module to sell a seat as this is a frequently occurring sequence.

If your firm assigns manufacturing part numbers in blocks, then the appearance of the first part number in an unassigned

System Analysis   79

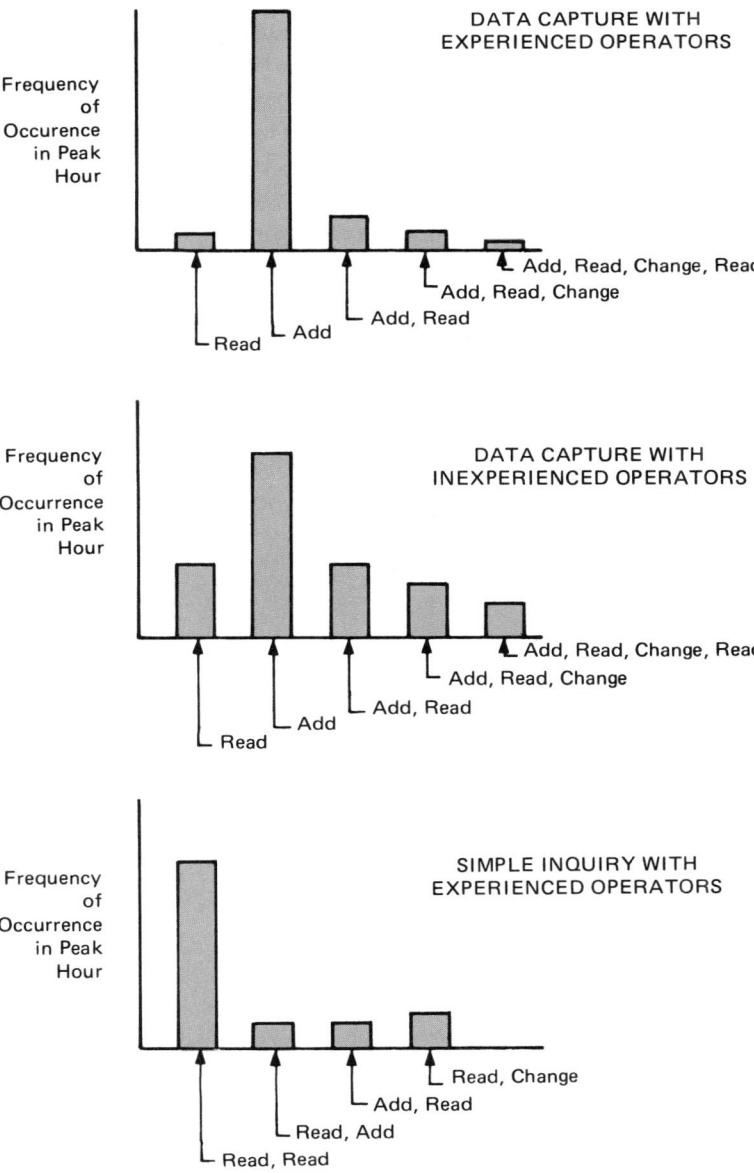

FIGURE 4.3   **Transaction Sequence Histograms**

block usually heralds the sequential assignment of all part numbers used to make up a subassembly. This is the way engineering departments traditionally release designs.

Figure 4.3 shows some histograms which depict this phenomenon.

73. Transaction-Data Element Matrix

V. Leontieff, in his Nobel Prize winning work, developed a technique for economic analysis called the Input-Output Matrix. (This is the same matrix that forms the backbone of linear programming models.) This analysis technique has been adapted for file design.

If transactions are listed horizontally as rows of the matrix and individual data elements from the data base appear as columns in the matrix, then an X can be placed at the intersection of a row and a column to indicate that a given transaction type requires access to a specific data element at processing time.

For small applications such a matrix is frequently constructed. For larger applications the technique is applied, but the entire matrix, which might involve 200 transactions and 500 data elements, is seldom constructed in this most detailed form. However, picture the detailed matrix as you read the remainder of this discussion.

Just as is done in mathematics, where pairs of rows or pairs of columns can be exchanged without changing the meaning of the matrix, a designer can interchange rows and columns of the transaction matrix to cluster transactions that are logically related, and to group data elements that have similar access patterns. After the data elements are grouped, some elements are always left over that do not fit naturally into any group. Using transaction sequences described in the previous item, one can see how many unattached data elements can logically be attached to a group of elements, so the probability of extra redundant file accesses is minimized.

The natural data element groupings are logical candidates for physical data base segments. However, before segments can be defined, a file key field must be identified. This is easily done by building a companion matrix which displays the data elements present in each transaction. The transactions should be listed vertically as rows of the matrix, and the data elements present in those transactions are listed horizontally as columns of the matrix. As before, an X indicates what data elements appear in what transactions. This matrix can also be manipulated by interchanging pairs of rows and columns so data elements common to every transaction lie to the left of the matrix, and optional data elements appear to the right.

By simultaneously matching the matrix that displays transaction contents and the matrix that maps transactions to stored data, common data elements can be identified. These common data elements are candidates for file keys.

The analyst should seek out common data elements which have unique properties: names, part numbers, and personal identifiers. Thus, if a field like Social Security Number is available in most input transactions involving individuals, and if SSN is being held in the file, it is a logical candidate for the primary file key. If no unique identifiers appear as common elements in both matrices, then additional steps are required.

One step would involve constructing unique identifiers by assigning a sequential transaction number to every transaction as it is received, or using forms which have been prenumbered to create unique transaction identifiers. In some cases the source of the transaction plus the day and time of receipt will be sufficient. If the unique identifier is a universal attribute of the article or person being described in the transaction, then it is likely to be the file key; i.e., the part number or Social Security Number.

However, if the unique transaction numbers have no intrinsic identifying properties, then it may be necessary to use some of the data provided in the transaction to perform a file look-up so a unique identifier can be obtained. For example, given a make, model, and engine serial number, the vehicle license number can be found. As a last resort, a unique identifier can be generated by combining certain fields of commonly available data, much as the file access methods of the 1960s scrambled data fields to produce a physical file address. For example, person name and date of birth can be combined to provide an identifier which is sufficiently unique for most purposes.

All this complexity is not usually required. Part numbers are unique and usually constitute the subject of the message; Social Security Number or employee number are unique when dealing with personnel files. Fortunately when the two matrices are compared, there is frequently a unique field on the transactions that will suffice to be the natural file key.

Recently concern for personal privacy and industrial confidentiality have added a review step to the process of logical file design. Sometimes records contain fields with different degrees of sensitivity and hence the system must provide two levels of authorization control. In these cases it is often necessary to split what would be a single logical record into two related records with a common key. One set of records would be stored in one data set with one set of authorization rules, and the other set could be stored in a second data set with a different set of authorization rules. Thus the need for security and privacy may disturb natural file groupings.

After the data elements are combined into natural groups, and a key is identified which is common both to the stored data and the transactions which access it, the transaction frequency histogram of Item 71 should be reviewed. Even though many data elements appear to have similar access patterns, if one transaction has enough volume so that it dominates the whole process, revisions to the evolving logical file design may be in order. The high volume transactions then can access data records which are compact and hence can be processed in the most efficient manner. Although the infrequently occurring transactions might be required to perform an additional file access, the overall efficiency of the application would be improved by a design which favored the dominant transactions.

74. Transaction Processing Time

Even while doing a logical file design, a designer should be aware of constraints imposed upon him by the hardware and software. The previous three items have noted that different input transactions require different resources for their processing. For data capture a single input transaction may result in one or two disk accesses. On the other hand, an update may require six or more disk accesses depending upon whether secondary indexes must be accessed and how many levels of secondary indexes must be traversed before a unique file key is obtained.

Similarly not all reads require the same number of disk accesses. If the physical sequence of the file can be guaranteed to match the logical sequence of the file, then the direct access path to every item is the same. However, if data has been added to the file in clusters and then one of those added records is needed, the record will quite likely be located in an overflow area and hence two or three additional disk accesses will be required for retrieval.

Thus the designer must be familiar with the concepts behind the file access methods he plans to use so the proper weight factors can be applied to estimate resource usage and processing time.

75. Analysis for Distribution

Analysis matrices are also useful in recording the structure of a system. Consider the following two matrices:

The left matrix maps the transactions to the modules required for their processing. If utility processes such as dispatchers and queueing modules are ignored, then the pattern of Xs in the matrix will indicate how many different transaction types require each process module. A few Xs in each row are typical of systems that are loosely coupled. Many Xs in each row indicate a growing complexity that should be analyzed in further depth.

Similarly, a report is usually generated by a single process. If a report is generated by two or more processes, it should be reviewed to see if its content consists of lines of some standard status report being generated in multiple modules, or if a new module should be defined so the generation of a report which occurs in three or more modules can be centralized for easier maintenance. If these two matrices are produced at design time and retained to be part of the maintenance documentation, service programmers will be able to assimilate the system and successfully make minor changes with less effort.

Two other matrices shed some light on system distribution:

If a matrix is prepared showing the relationship of transactions to the locations where those transactions are originated, and a second matrix is prepared showing the relationship of reports to the locations where those reports are used, some additional insight into systems design is obtained. If status or inventory usage data are entered at outlying locations and if corrections and reversing transactions were entered by a central group, this pattern would be clearly shown since the location matrix could be easily subset. In contrast, if any transaction can originate from any location, the matrix would be solidly occupied by Xs. If the reports matrix showed every location needed every report, one might consider printing only a reference hard copy of each report and holding the report on-line for terminal inquiry. The more usual pattern will show a report being used for action at one location and for reference at another location. In this case a designer should consider hard copy printers at the locations where the reports were used for action.

Analysis matrices can be enhanced to provide more information. If instead of Xs, letter codes are used, one can indicate where a report is used for reference and where the action takes place. In the case of transactions and their data elements, key fields, mandatory fields, and optional fields can be illustrated by one-letter codes. Or in the case of processes versus elements from the data base, one can indicate where values are created, read, modified, or deleted. Sometimes the element space in analysis matrices is split so two numbers may be entered at the intersection of each row and column. In these cases the function being performed is coded in one half of the space, and the number of times it is performed is entered in the other half of the space.

Thus the matrices have several useful properties. First, the pattern of entries illustrates the logical connections; second, the function codes describe the activity being performed; and third, the frequencies note how often those functions are performed so intelligent design decisions can be made with performance in mind.

When dealing with transaction matrices, a comments column is sometimes added so any transaction dealing with the file key can contain an annotation describing the range of keys considered. In some applications this column adds no additional insight. However, in other applications one finds some physical property of the application translates directly into a restricted key range. Consider the case of a manufacturing inventory system where one plant made engines and another plant made accessories. The part number (key) ranges produced by the transactions at these two locations might not intersect. Thus only the central office, where orders are received, and the assembly plant, where complete vehicles are assembled, would need access to the full range of keys. Any time such a restricted range of keys is discovered, it should be carefully reviewed as it may define a method for logically segmenting the data base.

When designing a major application which is to be distributed across several sites, all analysis matrices described in this and the previous items are frequently prepared. Furthermore, a set of matrices, identical in format, is usually prepared for each site; and then a composite set of matrices is prepared for the system as a whole. One often finds that transactions, or transaction-key range combinations, are unique to some sites. Even when transactions, processes, and reports are required at multiple sites, the activity patterns and the volumes involved clearly indicate where the data should reside and in how many places each processing module must be cataloged for execution.

If a set of analysis matrices shows a clear-cut pattern, a designer is tempted to conclude that the matrices dictate how the processes should be broken up and how the file should be allocated. While this may be true, a word of caution is in order. When dealing with analysis matrices, one is tempted to act as though all data elements are the same length and all process modules require the same number of cycles for execution. While the data elements may be homogeneous in length and the processes may be uniform, this is not usually the case. If a designer were using these design techniques to determine how to distribute processing now running on a host computer, he might elect to leave the long running processes, and the data they require, on the host and then attempt to distribute the rest. One may not always be successful in such an attempt, however. The data elements necessary for the compute-bound process may lead inexorably into a centralized master data base with portions replicated at remote sites supporting responsive terminal applications.

76. Design Documentation

Given a small application and no turnover in the design team, informal documentation will produce a satisfactory design and may be adequate for operational maintenance. However, if the system is large or if the maintainers will not be drawn from the design team, more formal documentation will be required for operational maintenance. If extensive documentation will be required, the earlier it is produced the more benefits are reaped. Formal documentation is useful if turnover is expected among designers, for use by programmers building test cases, for instructors planning training programs and, of course, for operational maintenance. The documentation prepared for distributed systems should include:

A. A dictionary of data element definitions.

B. Cross-reference lists combining data elements into groups, groups into records, and records into data sets.

C. Cross-reference lists relating data elements (groups, records, and data sets) and program modules.

D. Activity lists for each program module which show whether the program reads, adds, changes, or deletes the data elements it references. (These may be annotations on the program versus data element matrix.)
E. Location lists (or matrices) showing where each data element resides and whether the copy is the master copy or a redundant value.
F. Although detailed lists of personnel and their access privileges will not be required until the system becomes operational, the designer must consider access lists early in the design cycle. These affect the packaging of the transaction processing modules into programs and the definition of data sets that have uniform access requirements. Thus designers must provide for security lists which link:
1. People to programs.
2. People to files.
3. People to commands.

77. Data Base Design

As can be inferred from the previous discussions, external interfaces to a distributed application are rather easy to determine. Perform a frequency analysis on transactions and reports and you will find out where transactions originate, where reports are required, and the frequency of occurrence for each. Splitting up the data base is not nearly so straightforward.

As has been noted, the running time of some program modules may dominate and this in turn may dictate where some of the data must be stored. In some cases the activity patterns set ranges on primary file keys, allowing a single logical data base to be subset so the data records are stored near the point where transactions originate. In other cases requirements for security dictate that logical records must be split and held separately. File level access privileges will then provide a person access to only the data elements to which he is entitled. If any one of these considerations sufficiently defines rules for distributing the data base, then the reader can skip the remainder of this item.

However, many designers will find that their search for the optimum data base design is more complex than these simple cases. At the outset remember that data stored on the processor to which the terminal is connected provides the best response. Further, if the matrix relating transactions to stored data and the matrix relating originating locations to transactions both show a single logical path from one set of terminals to one group of records, then there is no need to replicate data or to provide remote access paths from other processors. However, other nodes frequently need to access data which is "owned" by some other node and central host systems or central nodes frequently require summary data from all other nodes in the network. In these cases the designer must consider replicating data to provide performance, and then providing some mechanism to keep all the replicated data elements current, or to provide only a single copy of any one unique data value and route the transactions from their point of occurrence to the node that owns the referenced data.

Detail analysis will sometimes show the existence of some transactions which reference data owned by two or more nodes or transactions that affect both detail and summary data. To accommodate these cases the system must be designed so a primary transaction can generate secondary transactions that request data from an adjacent node, or transmit summary data to the host in a hierarchical network.

If the files are small and the volume of transactions low, the designer will get the most responsive system by replicating the files at each location and propagating the updates to all nodes holding a copy of the file. The designer should also build some controls so terminal operators need not wait for a positive response unless the transaction caused an inventory to become alarmingly low or some other noteworthy event had occurred. Note that if the file is large or if the volume of transactions was great, the cost in storage or communication lines may be significant. A designer then should consider some options for splitting the file rather than replicating it as proposed.

If the file can be split horizontally based on the ranges of keys to be accessed, then a single logical data base can be defined and appropriate subsets assigned to each node. This favors transaction processing, while making summary processing more difficult, since a single copy of the file does not exist at any one physical location.

Some applications and the physical location of the personnel they support favor a data base which is functionally split. Consider a planning organization which answers to a corporate central office where the financial books of accounts are kept and which is physically separate from a manufacturing plant. In this case there would be three major data sets, all covering the full key range, but which support (a) planning and future build schedules, (b) status of work in process, and (c) cost accounting. If the users were physically separated, then weekly build schedules could be sent from planning to manufacturing and shipment information could be sent from manufacturing to finance. If the payroll were run separately, the finance department would already have access to the previous week's labor distribution data. Thus the detailed transactions would be primarily local to the node which held the data and internode communications would be primarily unidirectional from planning to manufacturing to finance and back to planning.

While each of these primary data distributions exists in real life, there are also many complex combinations. Consider the case of the manufacturing organization which had an accessories division and an engine division, but had some accessories physically manufactured at the engine plant. Or consider the multinational corporation which had one set of consolidated accounts, but found that foreign currency restrictions caused a duplicate data base to be held in each foreign subsidiary so that profit and taxes could be properly calculated. In these cases the communication links must be sized so detail transactions can flow between the sites. The designer must determine precisely the user's need for time currency to determine whether these detailed transactions must be simultaneously reflected in both copies of the data, or whether they can be batched at the originating location and transmitted to the adjacent site at such times and under such conditions that they do not add to the peak communications or processing loads.

Several considerations are omitted from most articles that appear in the popular press. Unfortunately the items overlooked involve some of the more important design considerations. If all data of a given type and all processing that requires that data is isolated to a single node, then the problems of data administration, changes to the data element directory, and changes to the set of processing modules are contained and rather easily managed. However, even if the data and the processing are unique to a node, if transactions are entered at other nodes and routed for remote processing, the application still involves multiple transaction definitions at the several nodes and transaction routing tables, all of which must be simultaneously maintained when a change is made.

If the data base is replicated at several nodes and if one set of processing programs is prepared centrally and shipped to each of those nodes, then central system maintenance is rather straightforward provided the system is not so huge that it prohibits the downline loading of changed modules to all locations "simultaneously."

If the data base is split so records within specific key ranges are assigned to specific nodes and if a key range table is used for transaction routing, the system has no redundant data but processing modules are cataloged for execution at each of several nodes. If the system is properly designed, it will measure and record inter-node traffic so the system manager may recognize changes in activity patterns. Consider the case mentioned earlier where accessories are manufactured in the engine plant, and hence part numbers for those accessories are in the data base at the engine plant. If the engine plant expands and the accessory operation is physically moved to another location, then it may be necessary to migrate the data to another node and again change the transaction routing tables in all the nodes.

Or take the case where the planning data, the manufacturing status data, and financial data were held in three separate nodes. The definitions of common data elements would have to be centrally administered and if need ever came to change them, all three data bases would need to be modified simultaneously. However, given that key or other identifying information was stable, the functionally split data base could be modified and changed almost unilaterally by the owning location since its unique data elements would not appear at any other location. Similarly, the processing modules would be cataloged at only one location since that is the only location with the data. However, the procedures for introducing new part numbers into the active file or retiring part numbers from the active file are somewhat more complex since the three files must be updated in the proper sequence.

Restart and recovery is another matter requiring considerable attention. In the case of the full data base that resides at two or more locations, any transactions entered at an adjacent node and routed to the node owning the data base must be retained at the originating node until their receipt has been acknowledged by the receiving node. Some designers maintain a queue of transactions in the originating node and require the node performing the processing to send a message that closes out that transaction after it has been reflected in the data base.

If the data base has been replicated at several locations and if the natural activity pattern allows processing to catch up and all deferred queues to be exhausted during a quiet portion of each day, then file control procedures can be programmed to assure, at least once a day, that all transactions in the system have been processed and that all of the replicated data bases are in fact identical. Once this has occurred and all item counts and control totals have been reconciled, summary data can be drawn from the system and management reports can be prepared. However, if the flow of transactions is continuous and quiet periods do not occur, then reconciliation must be dynamic. Controls over the processing of queues must be established so that coherent data is used for management reporting.

While on-line transaction driven systems are popular because of the response they provide to the terminal user, the designer cannot neglect to consider the needs of financial officers or governmental subdivisions for summary reports on a calendar basis.

As mentioned above, if the system enjoys a few hours of "quiet time" when no transactions are being originated, then file reconciliation and financial reporting are rather straightforward provided the active portion of the file can be processed in the quiet time available. However, with very large systems or very complex systems or systems which must be constantly available for a continuing flow of transactions, the process of "closing the books" requires some thought. Introducing additional data elements such as date and time of last update into the file may be necessary so duplicate records may be created while the file is closed for balancing. Then when the file is reopened for routine activity, the older records can be deleted.

Finally the problem of file backup, catastrophe recovery, and archive files must be addressed. Even though modern computing equipment is very reliable, some business managers are understandably reluctant to trust the only copy of their file to a single set of hardware and software for an extended period. This reluctance is one reason for full file duplication. If the file set is not overly large and if the system has a period of quiet time which is sufficient for a file to be copied, then a duplicate of the data base can be dumped to magnetic tape. On the other hand, if the file set is large or if the period of quiet time is too small, file copying must be a background task behind on-line processing. This will require additional programming so the copy of the file and the status of the queues are interlocked. Thus if the file is to be copied as of midnight on a given day, all transactions received prior to midnight should be reflected in the file before it is copied. If this is not possible, then one alternative is to copy the file as it stands and then to prepare a series of modification records for those transactions which arrived before midnight but were not processed until after the file was duplicated. If the system cannot be shut down to copy the file, then this bookkeeping must occur dynamically to avoid the possibility that a transaction could be lost or duplicated in the event the file was ever reloaded following a catastrophe.

In handling really big files, or if a relatively large file is being transferred to a relatively slow output device, it may be necessary to maintain a log of changes made to the master file which are then transmitted to the host for updating on more capable equipment. Thus if the file were big and the time available for backup were short, the log tape could be processed on a host system in batch mode to yield a current copy of the data base without having to dump the entire data base. Then if a catastrophe occurred, the tape corresponding to the current copy could be positioned for a complete or partial file reload so recovery could be accomplished.

Note that if the same transactions are used to update the on-line file and then later used to update the backup copy, special controls will be needed to guarantee that these two files reconcile to one another. Such techniques are of particular interest to a designer contemplating a hierarchically distributed system where the outboard nodes contain unique data, but the configurations do not provide for ready backup on a duplicate disk or magnetic tape, and the communication lines are sized to the normal traffic so insufficient bandwidth exists for a full file dump or a full file restore in a reasonable time.

78. Table Driven Code

If your applications architecture calls for cataloging the same program module for execution at one or more locations, the module should be internally structured so the parameters which customize it for each location are isolated and placed in a single table. Thus, the processing code remains the same and the table provides the control parameters which properly route transactions, apply appropriate limits to allow editing of keys to see if they match the local data base, etc. If the table is separately structured, an application utility program will be required to set, display, and change the parameter table.

If the table gets too big, it can affect performance. In this case retain the concept, but break the table into pieces and locate each portion of the table next to the pages of processing code that reference those parameters.

79. Throughput Estimation

In many small computer systems, the interaction between the processor and the disk file controls overall system throughput. Even at an early stage in the design, throughput can be estimated by counting the seeks necessary to access the file for each transaction type and by scaling the transaction histogram. This gives the disk activity for the transaction load which can be easily converted into a time measurement by applying a factor for average seek time. To this time one must add the time for batch programs and activity logging. Unless the histogram was expressly prepared to depict peak hour activity, one rule of thumb suggests doubling the average activity to get an estimate of the peak.

If one or more applications are running concurrently or if the processor also supports activities such as remote job entry, seek time must be calculated for these activities also. It has been repeatedly observed that a lightly loaded system gives good response, but a system with disk loaded beyond the 70% mark requires the seek calculations to be carefully done to avoid saturation .

It should be noted that the designer's approach to large host systems is different than his approach to small distributed systems. With the large systems it is extremely rare for a single application to tax the capacity of the installed configuration; therefore one can concentrate on getting the applications up and going and then tune the set. However, with small systems, it is not unusual for a single application to tax the capacity of the installed configuration; therefore designers of distributed systems must design with performance in mind.

80. Preliminary Operator's Manual

After a preliminary design is prepared, surprising benefits result if the design team next produces an operator's manual. If the operators review the manual while the detail design is proceeding, you will receive feedback from your community of users which allows human factors considerations to be inserted into the design early. In addition to a system overview, the operator's manual should cover:

- morning startup
- routine processing
- evening cleanup
- abnormal cyclic processing
- special runs
- recovery actions

Most designers recognize two modes of man-machine interaction; namely, the infrequently used function, and the routine high volume activity. If separate sets of standards are devised to govern each type of dialog, then the protocols for all the infrequently used commands will be the same. While the high volume, frequent activities may be different, they too will share a (different) common protocol. The goal should be to allow infrequent activities to be successfully performed with a minimum of error, while frequent activities are fast and easy to do.

As indicated in the outline given above, the early design should address abnormal cyclic processing such as the activities required to "close the books" monthly, quarterly, semiannually, etc. Similarly, some special runs should be anticipated so the preparation of budget status reports and/or the entry of new strategic plans can occur at any time, not just at the first of the year.

81. Built-in Training Mode

Preliminary design time is not too early for the designer to plan additional support and service functions which are relatively easy to include if considered early enough. For example, unless an installation was so huge that operator training can be delegated to a training team and conducted on a computer dedicated to that purpose, the training of replacement operators, or the retraining of operators transferred from section to section, and the retreading of trained operators who have been on leave or assigned elsewhere suggests the system include a training mode which will allow operators to familiarize themselves with the system without damaging live data files. If the log-on sequence provides for a training mode and if message processing programs are conditioned by that mode, operators can log-on for practice and reference a sample file for training purposes.

Similarly, system exercises can be easily prepared if a terminal can be logged on in a test mode. The terminal can use production program modules but have its outputs derived from, or its inputs compared to, a set of sample test files.

Built-in production statistics will allow a supervisor to be aware of the activities performed by remotely located operators who are not directly visible. Further, the sum of such activity statistics provides systems level statistics for workload reporting and capacity planning.

As was mentioned earlier, security must be considered from the very start since a given logical file may have to be split into two or more physical files to provide proper access restrictions for persons with differing privileges.

82. Built-in Statistics

If systems level workload statistics are desired, a designer should try to record total activity and the rate of that activity. For instance, a run-time parameter could control how often the activity statistics are sampled. If the parameter were set to one hour, then the counters would be reset to zero once each hour and activity would be accumulated over the set time period. At the end of each time period the totals could be optionally saved while they are being added to the previous totals accumulated for the day. This of course would require two areas for storing activity statistics, one for the detail counts and one for the accumulated sums.

As a minimum one might consider counting the number of transactions from each screen and counting the number of outbound messages returned to that screen. If one also counted disk accesses, messages to and from the communication lines, and lines printed, a good measure of overall activity would result.

If the lengths of the various processing queues were stored adjacent to these activity statistics, then a single command executed at the end of a preset time period could record the activity which transpired during that period and the lengths of the queues at the end of the period. In addition to being useful for reporting and planning purposes, this activity table would be useful whenever the system crashed, upon restart, and immediately prior to shutdown.

83. Built-in Flexibility

When installing a new application in a heretofore unautomated environment, extra design flexibility will be required for:

A. Specification changes as the customer learns.

B. Temporary modifications which may be necessary to help migrate from the old system to the new.

C. Temporary code which may be necessary to initialize counts and totals to year-to-date values, unless system cutover occurs at the exact beginning of a new fiscal year.

84. Support Center Features

Unless programming talent is available at each node where a processor is installed, it will be necessary to set up a central Network Support Center. If this is a possibility, it should be considered early in the design process as it affects message format and the dispatching logic within the processing nodes.

The activity statistics kept in the nodes should be available for readout by support center personnel. Further, each processor should maintain an indicator which declares whether it is waiting, in a test mode, or in the production mode. This indicator must also be available to remote support personnel.

In addition, any operator should be able to address a message to the support center, the support staff should be able to store a message which is transmitted to each terminal at sign-on time, and the support staff should be able to send a priority message to all active terminals which is displayed at the next breakpoint. Needless to say, these features must be implemented in the transaction router, in the message formats, and in the structure of the tables used for storing status information within each node.

85. Revised Cost Estimates

Shortly after the preliminary design begins to emerge, most designers find it necessary and desirable to refigure their development costs. It makes no difference whether the design is expressed in flow charts, HIPOs, or in pseudo-code. In the preliminary design stage this documentation is neither exhaustive nor complete. However, if it is necessary to reestimate costs, one way to proceed is as follows.

Given an understanding of the application and a preliminary design, an initial list of processing modules, display screens, printed reports, and data sets can be prepared. Given experience with analogous applications on a small system similar or identical to the one chosen, guidelines for module definitions can be established so all named modules are roughly the same size and complexity. The same body of experience will provide a factor which when multipled by the number of mainline application-specific direct processing modules, yields an estimate of the total number of modules required and hence covers special tools, restart programs, file editors, and the like.

If the shop has experience with the chosen system and its software, then the labor records from previous jobs can be distilled to obtain estimating coefficients for effort, machine time, and elapsed time which when multiplied by the number of modules, yields project cost and schedule.

Such estimating is very preliminary and contains a great deal of uncertainty. Some of that uncertainty can be removed if the project environment from the previous project is compared with the project environment for the current project: that is, was it a rush job working against a deadline, or a more leisurely development which took its natural course. Similarly, the previous development team can be calibrated against the development team about to undertake the new job: that is, is one a seasoned crew of development veterans accustomed to working together, and the other a pickup crew of unknown individuals? For each of these extreme cases, the estimating coefficients will need to be adjusted accordingly.

One may reasonably ask "How do I make estimates if this is my first job on a system I am unfamiliar with?" Obviously one can estimate as described in the previous paragraph if one is willing to tolerate more uncertainty in the resulting estimate and schedule. However, a prudent person may not wish to tolerate so much uncertainty when installing a new application on new software and new hardware. In this case it may be desirable to proceed as before and enumerate all of the modules, screens, reports, and data sets to be developed. Then after an intensive study of the software provided with the chosen system and the standards established for distributed applications in your installation, pick a medium complex item from each of the following four types and: design a screen, code a module, lay out a report, and design a data base. Convert those detailed designs into source code, get some time on a demonstration system, and check out the set. (You will get additional benefits to your user relations if the set of things you picked for this pilot exercise is also suitable for use as a demonstration after the exercise is complete.)

If careful records are kept during the pilot exercise to determine how much effort each step took, and if that effort is subdivided to breakout the initial learning experience from the expected effort a trained person would invest in doing the same job, the resulting coefficients will allow rough estimates of the design, coding, and unit test stages of development. If no better estimating factors are available, these numbers can be doubled to cover training, documentation, system test, and installation, thus producing a gross estimate of the development expense *provided* no major manual master files require cleanup, keyboarding, and editing prior to system operation.

If your installation has experience with some other small computer, much of that experience will be directly applicable to a system from any other vendor. However a recalibration will be necessary before the estimating coefficients from some other hardware/software/application combination can be applied to a system from a different vendor. Depending on the features the software contains, the estimating factors can vary even when equally skilled and mature persons are suitably trained and experienced. One should be careful not to use coefficients derived from a mature development team, thoroughly familiar with the ins and outs of the hardware/software packages for some other small machine, to estimate the performance of a new development team preparing their first application on a new set of hardware and software.

86. Benefit Estimates

After the costs are determined, it will be necessary to estimate the benefits before the application can be considered justified. Given a preliminary design, prepare a list of application features in a difference list format (similar to Appendix B) so the functions which have a direct analog are described as a before-and-after pair; the new functions offered by the new system are crisply enumerated so their benefits can be weighed; and any old functions not planned for implementation can be highlighted for discussion. Such a difference list also sets the stage for a discussion of the tangible and intangible benefits of the new distributed application.

Most of the hard tangible benefits will be related to savings in personnel and improvements in process efficiency, while additional intangible benefits can be derived from improved data quality. The personnel savings must, of course, come from whole persons displaced and headcount reduced. It is insufficient merely to save time on an existing staff unless the workers in the unit draw frequent overtime or unless part-time personnel are employed. If a large group of people work in the same physical location, and if they perform the same tasks all day, then a 10% improvement in efficiency translates into a 10% reduction in headcount. However, such is frequently not the case and jobs must be restructured and persons must be reassigned before the headcount can be reduced.

Unless job restructuring is done carefully with full support of management and with adequate consideration for transferring or retraining the workers involved, the savings will not occur unless the work unit faces a growing workload and hence the new application allows the present staff to accomplish a greater volume with the same headcount.

In passing it must be noted that new computer applications are frequently received with a great deal of apprehension since workers feel (rightly) threatened by automation. Care taken during the design stage will allow the small computer to be used as an instrument of change *provided* care is taken and workers are assured of continuing employment elsewhere in the corporation after suitable retraining. Otherwise, if the computer is received as a surprise, the promised benefits will not be forthcoming.

The introduction to this item did not mention the saving of time as a benefit. This is because time savings must be translated into process efficiency before tangible results occur. Getting data faster yields no tangible benefit unless action is taken on that data. However, if engineering changes are transmitted rapidly to production coordinators, manufacturing process volumes can be adjusted so there is less scrap and rework following an engineering change. Similarly the availability of timely information may allow bulk discounts to be realized from a single purchase rather than two procurements each involving half the volume. Or the savings in time may be translated into additional production on a high-volume manufacturing line which yields a lower unit cost than would some other short-run manufacturing process.

Improved data quality derived from on-line editing with immediate feedback to the terminal operator sometimes results in a significant tangible dollar savings. In almost every information system, a significant portion of the operating dollars is spent checking for errors and correcting errors once caught. If the existing system is not sufficiently robust, errors can creep into products delivered to consumers in the form of incorrect shipments or incorrect accounting. Subsequent costs are involved in processing returns, restoring serviceable items to inventory, and scrapping or discounting items which are damaged. In addition to postage and handling, debates over bills and loss of customer goodwill further tax a business enterprise.

Industrial engineering units or customer service sections can provide statistics on the errors the customer sees and their direct costs. Labor distribution records can frequently cast light on the in-house efforts spent in catching errors or recovering from them. Given an estimate of the labor, a good cost accountant can estimate the other expenses incurred in each support function. When totaled up, these items frequently justify an on-line application based on improved data quality alone.

87. Cost-benefit Summary

After the development expense and the projected benefits are estimated as described in the previous two items, a cost-benefit presentation can be prepared. If the tangible benefits are substantial and the development expense nominal, a work sheet similar to Appendix A can be prepared which will yield a payoff curve similar to Figure 1.4. If the curve does not cross the axis in a reasonable time period (where "reasonable" is a function of the expected return within your specific industry), the intangibles depicted by the difference list of functions will need to be assessed.

If the decision is still too close to call, senior management may need to be reminded of the escalating cost of labor and the decreasing costs of computer technology to see if these trends change their opinions. Sometimes the scarcity of trained personnel or the expected growth in the business tip the scales towards automation.

Finally, only management can determine whether a competitive advantage now or in the future can be derived from having a distributed system in a marginal case and whether that advantage is sufficient to warrant proceeding through the design phase so more accurate estimates of costs and benefits can be obtained.

DESIGN

88. Roles and Missions

If the costs of rework are to be avoided in future years as an application grows, as corporate business units are physically rearranged, or as the product mix changes, the application designer must force the resolution of some policy issues involving the distributed computing environment.

If small computers are to be installed throughout a corporation and if decentralized development is encouraged without centralized programming standards being set, the resulting programs will be unique to each site. Even where the same requirements occur at two or more locations, it will be difficult to utilize applications that are developed elsewhere in the corporation.

Terminal protocols will be defined differently and even if local standards have been set, the standards from two sites will be incompatible. This will distract operators and increase both training time and error experience. Without standards, detailed data element definitions are unlikely to be common much beyond the Social Security Number/Part Number stage. Thus reporting of quantities-on-hand or calculations of earned-value on work-in-process will be different, and these differences will be hard to reconcile. Further, documentation may be skimpy or nonexistent and the applications may not provide sufficient information for troubleshooting in the absence of the designer. For these reasons, and many others like them, computer center managers have traditionally elected to recode applications rather than trying to utilize applications from another location which were designed without portability in mind.

Some of the deficiencies can be mitigated if the data bases and the applications are retained only at the site where the applications were initially programmed and if transactions from the second site are shipped to the initial site for processing. However, it is still likely that problems will occur since software may not be generalized for operations at two or more sites, troubleshooting may be difficult, and operating statistics may be hard to break out by site.

Another policy option, this one more practical, involves centralized development standards and decentralized developments. Thus development teams can be placed where the job knowledge is greatest, and the results of those development efforts can be reviewed by a central group prior to cataloging the programs in the central production library. Thus the central review group would set and maintain the standards, receive completed programs for review, and would review and test those programs just as if they had been packages purchased on the outside.

With the aid of a data dictionary, a set of standards, and a library of test cases, the central staff can assure that the programs are portable and that common data elements are commonly defined. Further, the standards will assure that the terminal protocols and the generic report formats are consistent throughout the entire network so the programs present the same "face" to the user regardless of the development location.

A third viable environment involves centralized development and decentralized maintenance. Thus each node would have a small cadre of technicians/administrators who could oversee system operations, set local system parameters, tune the system to its workload and configuration, and be the first line of defense in problem determination and applications repair. One of the problems with central development and service is the lack of response (both real and imagined) to the needs of a local site. Different time zones may be involved or in the case of multinational corporations, different nationalities. Surely there will be different work priorities since the goals of the work unit reflect the goals of the local site.

However, if central development and local maintenance are contemplated, not only will you need the standards and reviews discussed in the previous paragraph, but it will be necessary to prepare diagnostic and repair tools for use by each cadre at each node. The overall plan must encompass ways for the persons assigned to the cadre to be trained, to maintain their competence, and to be constrained so their developments and improvements are not incompatible with the efforts of the central development shop.

Last but not least comes a policy environment which provides for centralized development, centralized maintenance, but in turn implies remote diagnosis and fix. Many companies plan their networks around a central support staff. If high availability is required, this staff provides coverage programmers for every shift where availability is critical. By concentrating these talents in a central site, the maintenance of standards and quality can almost be naturally assured. Further, with a properly designed set of diagnosis tools, most of the maintenance can be done remotely. Then when help is required at a field site, it can be obtained from the vendor's field forces to repair malfunctioning hardware; or from the site administrator who will assist, by terminal and telephone, with that small percentage of problems which requires a combination of on-site and remote talent for their resolution.

While the choice of an operational environment is not an irrevocable one, the issue is fundamental and will affect the programming standards, the development methods, and the design of specific applications. An on-line dump on a high speed printer is meaningless if the programmer who needs that dump is a thousand miles away and if the volume of the material to be printed saturates a low speed communications line. Some vendors' basic hardware and software are designed with remote problem determination, remote software service, and on-site hardware maintenance in mind. To fully realize the benefits of this design, the applications must be similarly designed so they support remote problem determination and maintenance. This implies the need for keeping status and statistics in both electronic and hard copy form, preparing software and procedures to allow applications to be diagnosed and repaired remotely, and installing other procedures to allow the transmission and distribution of necessary changes to documentation and training materials. There are many ways an application designed for remote maintenance differs from an application unique to a single node. Most of these differences will be enumerated in the items that follow in this handbook.

89. Specific Design Goals

When designing an application for a large host machine, one is seldom concerned with global design goals because these are usually set by the computer center director and executive management. However, when designing distributed applications, one frequently encounters problems establishing a new computer facility for the application to run on. In addition, the operational environment being established affects the application and in turn is affected by any application which is large enough to dominate the configuration. Thus designers of applications for small computers frequently must define and obtain definitive decisions from management in order to establish meaningful design goals for reliability, flexibility, accountability/integrity, and cost-effective growth.

While modern computing equipment is very reliable, it can fail due to hardware, software, or applications code. The designer must determine the user's degree of dependence on this equipment and seek the best way to satisfy the needs of the work unit even though the system may be unavailable. If the site is located some distance from the vendor's nearest office, it may take several hours for vendor field personnel to arrive after the service call is placed. A designer must ascertain typical vendor response times and weigh these against the needs of the work unit the application will serve. In many cases the hardware can be restarted, the failing unit can be unplugged, or otherwise made unavailable. If sufficient flexibility is designed into the configuration, alternate communications paths will allow a user to enter critical data from an adjacent terminal, printed output to be displayed (in the event the printer was down), or transactions to be queued in the event a portion of the file was unavailable.

If the application was to be installed in many locations, then it must be designed with sufficient flexibility to allow it to operate satisfactorily on a variety of configurations facing a variety of workloads. If it is absolutely mandatory that the system retain all records even when faced with a variety of potential catastrophes, then the file system and the activity logs must be designed to survive. Duplicate copies of the file, duplicate copies of logs assigned to different physical disk drives, or backup copies of logs stored on magnetic tape or transmitted to an adjacent system over a communication line may provide the appropriate insurance to guarantee file integrity. If a system is embedded in a larger operation through which paper flows and is stored and retained in an archive, then the loss of a file record need not be so critical if the record can be identified and the manual paper handling system can provide a copy which allows the record to be re-created.

If a series of branch offices all face the same problems but vary in size and workload by a factor of four or more, the system — consisting of hardware, software, applications, and procedures — would have to be designed so it could be configured for the appropriate spectrum of workloads. The minimal system might consist of a processor, a few terminals, a disk drive, a hard copy printer, and a communication line so data could be stored off-line for backup. As the configuration grew and more terminals and disks were added, it could accommodate more workload. In yet a larger installation more terminals, a second printer, and a magnetic tape might be installed. If the workload were great enough, two systems sitting side by side with a split data base and intersystem message routing might be required.

If a user expects to maintain one basic copy of the applications code and to have it operate efficiently across such a wide spectrum of configurations, this requirement must be factored into the design or the program structure will not allow the additional hardware to be properly exploited.

Some systems cannot be shut down for maintenance. If 24-hour operation is required, at least two interconnected processors must be installed, software extensions must be planned, and the application programs must be designed so that two machines can carry the peak load while either machine can carry the off-peak load. Then either machine can be maintained, serviced, or used for program development. Designers will be challenged by the need to plan and design file systems which will allow new data elements and revised formats to be reflected in the data base while the system continues to operate.

After an applications architecture is determined as described above, a designer must test that architecture for the abnormal environments it may face. Not only must maintenance, service, reconfiguration, and degraded operation be considered; designers must also concern themselves with peak loads, month-end closings, unusual events such as leap years, and certain types of special runs such as the extraction of a large quantity of data from the data base for a special report or the need for a mid-year transition to a revised budget. Unless carefully thought through, some of these unusual conditions may become impossible to handle through some quirk of the design.

90. Accommodating Error

When dealing with a network, the designer must estimate the susceptibility of each terminal, loop, processing node, and communication line to error. Such errors are a function of intrinsic equipment design and the environment in which it is installed. By collecting error statistics on existing systems, the designer can frequently identify error-prone environments so the new design does not collapse in the face of expected difficulties. Duplicating communication lines, using modems which adapt and reduce transmission speed as a function of error history, relocating files away from environments subject to earthquake or other disaster, running two loops through a facility so adjacent terminals can alternatively be connected to each loop, maintaining a duplicate copy of the files on a second processor, and using a "pass-through" mode to provide service in the event the file system on the local processor is unavailable are all techniques the application's designer can use to work around environmental limitations.

91. Living with Error

In a batch system we customarily abort the job step whenever an "uncorrectable" error occurs. By this means, human intervention is obtained for conditions which occur too infrequently to warrant programming. However, in a distributed on-line system, a new philosophy is required. At the risk of overstating the obvious, an on-line system should never abort. Therefore, some of those infrequently occurring circumstances must be accommodated by programming. The remainder must be recognized and set aside in some meaningful fashion so human intervention can be requested. However, while the person is solving that problem, the system should continue to process transactions not related to the troublesome case and must queue all transactions which are logically dependent upon the case being investigated.

In the rare case of a nonrecoverable system error, the system still cannot abort but must do an orderly shutdown or chaos will result. For instance, if log messages are being blocked in a buffer, a controlled shutdown will cause the buffer to be written onto the storage device. A long running background application should be programmed so it will respond to a request for orderly shutdown and clear its queues, write end-of-file, and take a checkpoint so it can be restored without loss of the computing performed to date. This is particularly important if a background application were maintaining an on-line disk file, since backout of all changes made to the point of interruption and then rerun of the job from its beginning may be unreasonably slow tasks.

The complementary situation occurs in a large network when a system or node is restarted. Status indicators must be checked to see if all applications were completed prior to the shutdown or whether some application must be restarted partially through its processing. Priorities in communication messages were briefly mentioned some pages earlier in this handbook. While some networks require priority in messages to synchronize clocks, perform problem diagnosis, or start beginning-of-day activities, a system which contains application programs that are sufficiently long running to require orderly shutdown and restart in times of emergency, also requires message priority carried over into the application. Then the predecessor messages contained in the queues when the checkpoint was taken are processed before recently entered current transactions.

92. Operator-absent Operation

If the remote node is to be set up for operator-absent operation, it must be possible to restart the system remotely. Some vendors provide standard features to make this possible. However, to make operator-absent operation a reality, the application designer must build in code which responds to an upstream command to quiesce the system. Then status can be determined prior to reloading the system. The designer further carries the responsibility to build applications which can be remotely cold-started or warm-started depending on the conditions at the time the restart is required.

Remote recovery, file diagnosis and rebuild, and system diagnostics were so arduous in one large network that these programs were prepared to run on one of a pair of processors in the node while the other processor carried the current workload.

93. Trouble Indicators

System operators have long used response time, at any level of transaction volume, as an indicator of a system's health. Thus if terminal response time deviates significantly from the expected response time at a given level of transaction load, an alarm is sounded so the support staff can determine whether diagnostics are in order.

Another technique which is easier to implement (and is in some ways more informative) calls for threshold limit checks on all queues maintained by the application. These limit checks detect abnormal conditions and sound the alarm *before* the queue overflows or uncontrolled lockout occurs. The abnormal conditions are reported to the support center staff and processed by an abnormal condition module which is empowered to change priorities, shut off inputs, or take snapshots while the support staff is diagnosing the problem.

94. Flexible Designs

In a multiprocessing network, transaction loads frequently differ from site to site. If this variation is significant, configurations must vary to minimize equipment costs. When programming in this environment, device allocations should be contained in explicit address tables. The application code should be constructed so it will run on a minimum configuration and exploit the maximum configuration available. The code should be biased so it is efficient for the maximum configuration even though it may use more cycles on the smaller configurations where the cycles are more likely to be available.

95. High Availability Configurations

A popular high availability configuration is achieved by placing two interconnected processors at each site and connecting the terminals so every one is available to both computers. The computers would then be programmed so that one machine processed half the messages it received and logged the other half, while the other machine processed the complementary traffic. Then if either machine went down, the surviving processor could process its log file to pick up the remainder of the load.

96. Risk Analysis

When designing for high availability, many analysts list the problems they can foresee, assign a probability to each, order the list on the basis of probability of occurrence, and address all the problems above some cutoff probability level. They then treat the rest of the problems in bulk by trying to retain enough information for the system to be restarted with manual intervention.

Other analysts go a step further and assign a cost of loss to those problems destined for bulk treatment. In some cases the product of the probability and the cost is great enough to warrant programmed treatment.

Given such an analysis, the programmer can construct a cause and effect list to define the degradation that results from each problem. Once this is done, specific detailed analyses will uncover the symptoms of the problem, the problem recognition process, and the actions to take which will result in continued system operation in a degraded mode. However, the analysis should not stop there.

Once a system is in a degraded mode, repairs must be accomplished to the hardware, software, applications, or the data, and then normal operation must be restored. A story (which may be apocryphal) makes the point: a processor had two disk drives. The A drive held all the programs and data, and the B drive was used for backup. When the A drive went down, the processor switched to the B drive and continued operation without degradation. When the A drive came back up, the current queues were transferred from the B drive to the A drive so the normal configuration could be restored. One wonders why the system did not continue to run production from the B drive and use the A drive as backup.

97. Terminal Blackout Time

During a diagnosis or recovery process, the problem determination program sometimes seizes exclusive control over the data base. If the system is to be programmed so partial operation can be restored as soon as recovery is partly accomplished, then the diagnostic program should set a maximum lockout time parameter so it voluntarily relinquishes control to production modules periodically. Otherwise the terminal service on the restored portion of the system will suffer unnecessarily.

Maximum blackout time is the time period during which an operator's screen goes dark while the system is recovering. Each application and each environment differs in its tolerance for blackout. If the natural recovery time for the frequently occurring problems exceeds the user's blackout tolerance, then designers must seriously address the problem determination and recovery activities.

In large batch systems a multistep job occasionally must be restarted at the beginning of the job, losing a few hours of processing time. While this may be the simplest type of recovery to program, it is usually unacceptable in an on-line environment. Thus special programs are required to assess the status of the data base, and special data base structures are required which contain sufficient redundancy to support this assessment. The entire file system must be designed so the file set can be partially locked, allowing production operation on the surviving members while the records in error are reconstructed. Thus the maximum time of total blackout is reduced even though the partial blackout may apply to some records for an extended period of time.

98. Limited Compatibilty

A recent paper notes the need for a compatible interchange between unlike systems: source programs, communications, transactions, portable media, and data structures.* It should be noted that software incompatibilities can exist even though all equipment is obtained from the same manufacturer. Beyond that, incompatibilities can exist between different application programs since they may have been designed at different times to conform to different sets of local programming standards. If one expects to transfer data between incompatible peers in a network or between incompatible machines in a hierarchy, the data structures should be given meticulous attention. The state-of-the-art does not support logical subsets and extensible systems for the general case.

The further designers move from logical subsets which can be directly transmitted, the more translation code will be required to transform one set of record structures into the other. Even if these transforms are carefully designed and table driven, they will increase the processor load due to communications traffic and occasionally will inhibit problem determination and program change.

99. Message Routing

Addressing in a network requires some attention. If the same application is to be cataloged and executed on more than one node, it must be able to identify the physical processor it is operating on to construct internode messages.

Similarly if individuals transmit internode messages, a routing table is required. The return message may be routed to

---

*A. L. Scherr, "Distributed Data Processing," IBM Systems Journal 17, No. 4, 324-343 (1978).

either the individual composing the original message or the incumbent performing that function when the reply is received.

100. Accommodating Communications Error

Large integrated networks involving heavy communications traffic frequently install analog communications diagnostic instrumentation and employ a communications technician to deal with line troubles. The advent of the more sophisticated communications protocols, such as SNA, reduces the need for such communications diagnostic centers. Further, since some undiagnosed communications problems seem to improve with time, multiple speed modems have been developed which automatically transmit more slowly when difficulty is experienced.

In addition, many installations have physically installed a patch panel so the correspondence between lines and modems can be changed as part of the diagnostic procedure. If more adaptors are ordered for the computer than there are active lines, then the correspondence between modems and adaptors can also be changed to continue the diagnostic process one more step. This gives rise to the need for another routing table since there will no longer be a one-to-one correspondence between an off-site physical location and the machine address of that location.

When dealing with noisy communication circuits, some of this complexity cannot be avoided. However, simple antidotes for communications troubles should not be selected lightly, since the ability to change the addresses of physical locations implies that restart is more complex. Further, the existence of a telephone patch panel introduces a security risk which requires containment. Manual activities related to communication lines will require manual entries into the automated log, or the systems configuration audit trail will not be intact.

101. Support Center Messages

If a central network support center is contemplated, the handling of its message traffic deserves some attention. First, it will not be possible merely to transmit messages presented to the local operator, as the messages will not contain sufficient identification to be meaningful to the central staff. Further, the central staff will need the local context to interpret those messages intelligently.

One technique would be to interpret all messages going to the systems log at each local node and to transmit a subset of those messages to the central support center. Thus, in times of routine operation the subset transmitted would be quite nominal, consisting primarily of workload and status information plus information on all hard device errors and all reconfigurations. If the system were operating irregularly, the subset transmitted could be increased to include local context (job start and terminate, log-ons and log-offs, error messages, etc.) so the central staff would see all operator messages plus the context needed to interpret those messages. Finally there must be an option to format and transmit every message going to the local systems log.

As discussed previously, if a system failed without warning, the central service center could restart it remotely; scoop out, format, and transmit system status tables from the remote site to the support center; transmit a portion of the recent log to the support center; and probably would set the interpretive logging program to transmit all new messages until routine operations were successfully restored.

Options must be provided in vendor software to support operator-absent installations. In some vendor offerings a person at the central support center can sign onto a remote system and enjoy the full privileges of a local console operator. However, some confusion is possible in the event an operator is present and help is also being provided from the central support center. In this case the programming for the local log and audit trail must be careful to record which operator gave the command, since the node may treat locally generated and remotely generated commands in exactly the same manner.

A network support center routinely needs:

A. Traffic statistics
B. Error statistics
C. Component status

D. Log of actions and log of selected activity
E. Log of changes

The network support center also needs the ability to:

F. Load programs, tables, and patches
G. Start/proceed from the point of interrupt or start at some arbitrary address
H. Stop/save context or halt
I. Reconfigure switches and tables

The operations procedures at the central support center need to be set up to deal with the information routinely provided by each system so the status, context, and configuration are instantly available in the event problem determination assistance is required.

102. Network Management

The systems aspects of network management require attention during the design phase if the network is to operate satisfactorily during its productive years. The design team on a big network expects to spend some time in this activity, although many tend to underestimate the effort required. The design team installing a stand-alone computer system may wish to ignore this important area altogether and as a result may be foreclosing future options to hook up to an evolving network and operate satisfactorily. The symptoms of a bad hookup are occasional excessive down time and continuing excess costs for both maintenance and service.

To summarize some of the items which have been presented before, it must be noted that a successful network requires:

A. A master plan.
B. Programming standards.
C. A migration methodology.
D. Methods for keeping replicated files synchronized.

116 Design Phase

- E. Controls (several types) and audit trails.
- F. Service level reporting.
- G. Workload measurement and capacity planning.
- H. Several types of training.
- I. Provision for network problem management.
- J. A spectrum of diagnostic messages which address both hard failures and intermittent operation.
- K. Procedures to maintain the application code, the data bases, the related manual procedures, and their documentation.

103. Plan for Change

Planning for change is more critical in a distributed environment than in a batch environment. There are finite physical limits on the number of messages that can be transmitted over a communication line and the number of transactions that can be entered from a single terminal. If the business unit being supported fluctuates in size, this can cause reconfiguration of the computing system to accommodate growth or to reduce cost. An on-line system is much more closely coupled to the fluctuations of the business unit than is a batch system. If the business unit is split, then the computing system may need to be extended or replicated. If two systems are supporting a manufacturing operation and additional manufacturing facilities are built, terminal placement or communication lines may need adjustment even if the total volume remains the same. Thus computing system planning must be kept in step with the planning of the business unit. A plan for change should address the following items:

- A. Programming standards to provide table driven code, good documentation, and good diagnostics.
- B. Utility programs to create, display, and modify control tables.
- C. The ability to temporarily immobilize the system, a node, or one or more terminals and then change configuration tables or edit rules.

D. Provision for recording traffic measurements, service level measurements, and trouble measurements in machine-readable form. Then an analysis program (and some procedures) can process the data to predict/recognize problems.
E. Modular hardware, programs, data files, documentation, training, and controls that allow reconfiguration to be relatively easy and the resulting system to still be efficient.
F. Flexible computer hardware, communications lines, power, air-conditioning, and site layouts arranged so that the hardware and communications subsystems can be reconfigured as required.
G. Plans for an administrative change control system covering:
   1. Standards for design, programming, and testing.
   2. Production application libraries.
   3. Data base definitions and structures.
   4. Information exchange for node-unique problems.
   5. A change control procedure providing information on, and controlling the introduction of, changes.

104. System Decomposition

In a paper concerning communications and distributed systems, one observer* noted that the system must be decomposed to be distributed, and that good decompositions have two properties:

A. The parts are more homogeneous than the whole; and
B. The parts are weakly interconnected.

105. Designing for Change

Designers are frequently admonished to "design for change." Here are some rules that will help:

A. Separate editing from processing.

---

*Richard Van Slyke, Network Analysis Corporation.

B. Design edits so they are table driven and can be easily changed.

C. Condition the design of the edits by the errors the system will experience. Since the errors seen by the system when it is new (and all the operators are new and all the documentation is new) are different from the errors the system will see after it settles down (and operator skill increases), plan to enhance the edits as time passes. Plan to edit for stupid mistakes when the system is first installed but make the edit checks open-ended so they can be enhanced when more is known about the sophisticated mistakes that are made.

D. Design the data bases properly in the first place so they will contain the data necessary to run the business. If the business is stable, the data base tends to be stable. Therefore be sure your first data base design is a good one so the changes will be aimed at performance and not new function.

However, as insurance, keep a clean interface between the applications processing module and the data management system so one can change without affecting the other. In addition, always refer to data elements by name and let the compiler resolve the names into machine addresses. Then if it is necessary to modify a record or add a data field, a simple recompile will yield a running program that is compatible with the new format.

E. Assume that constants will change and rates will change faster. As a result, use a constant for only one purpose and resist the temptation to assign two meanings to the same constant even if the numerical value is the same.

Rates belong in a table. Tables should be a separate module. If tables change frequently, then an on-line table maintenance and display program will be required. If tables change infrequently, they can be manually edited and recompiled. Never build rates into a program as literal values. Further, unless it makes a major impact on performance, never put the same data value in twice; i.e., provide a single copy and make all references to that one copy. In the case that multiple copies of a constant or a rate table are absolutely required for performance, be sure to annotate both duplicated values so a programmer later changing a value is reminded, "This also exists in Table 3."

F. Be sure the reports and screen presentations are kept separate and distinct from the processing since the part of the system most likely to change is the output.

When preparing a batch report, plan to sort the data after it is selected from the file and before it is printed. Then if you do not need to sort the data, nullify the sort step. However, if someone later wants the report in a different sequence, the output module will already be structured so a working file can be written, sorted, and read in between the select and the print steps.

If display screens are laid out properly (good readability, consistency within the set, and containing appropriate data elements so a variety of questions can be answered), then the most likely change is the addition or deletion of a few data elements within the basic format provided.

G. If multiple files are involved but the data base is not integrated, expect a likely future change (one which is very hard to prepare for) requiring the addition of a new display screen which needs coordinated access to two or more files on some common element. This will give the illusion of an integrated data base, even though separate files may be held.

This is one of the series of changes that cannot be anticipated and will not be easy to make because the idea of an integrated data base was probably rejected in the first place for some nontrivial reason. That reason will come back to haunt the designer who tries to simulate coordinated file processing.

H. Finally, implement the programs in the set so they are relatively easy to change. Structured design and structured programming will provide that ease of change by making implementation straightforward and minimizing the number of intermodule connections. However, the programs will be a little bigger and run a little slower than a monolith which does not follow such rigid conventions. In anything but a real-time environment, the space and time for structuring the code is probably well spent and will reduce the maintenance costs over the life of the system.

106. On-line Update

Some conservative designers try to avoid on-line data base update if they can do so without sacrificing an important user requirement. Since their reasoning involves more than hardware reliability, it deserves serious consideration.

One of the principal advantages of an on-line system is to capture the data from the outside world as it is available. An on-line data capture application which builds a queue of transactions for later batch update against the data base provides that capability. Later after the workforce has gone home, the queue can be processed and reflected in the data base. In this example the data base would be current during the evening hours and through the morning until the first transaction of the day was received. Reports prepared during each day and inquiries received throughout the day would be pulled from the data base which was current to the close of the previous day's business. Thus the data base would be read-only throughout the day and updated in a batch every evening.

The advantages of this mode of processing are ease of programming and a greatly simplified restart. The problems of dynamic space allocation and update processing during data base recovery simply do not exist. The disadvantage, of course, is the fact that the data base does not instantaneously reflect the course of the day's business. However, unless one is supporting an airline reservation system, this may not make any difference. In a manufacturing environment the build schedules may be periodically set anyway, and the processing of a new order may not affect the outcome of a day's production in any event. In the case of payroll or personnel applications, the penalty for not doing an immediate update may mean that an incorrect name, address, or phone number merely lies in the file one additional day.

However, if you are faced with an application requiring files which are updated to the minute, you must be sure to design with recovery in mind.

107. Nonstop Operation

When designing a large application system that is started once in the morning and is intended to operate all day long in nonstop fashion, several hints come to mind:

- A. The principal parameters which control the configuration of the system at any point should be gathered into a table.

- B. After the computer hardware and software are initialized each morning, the initial applications module should provide the system administrator with the exclusive capability to configure these control tables to match the current situation and set up operations for the day. Thus the system can be brought up for part of the workforce on weekends without exposing all the files to mischief.

- C. Any statistics, measurements, or accounting information that is collected throughout the day should be periodically stored so all of these measurements are not lost in the event of a system malfunction. When the system is shut down at the end of the day, the final accounting records should be uniquely identified so these can take precedence over any intermediate records that were periodically placed on the log (for insurance purposes) throughout the day.

- D. When the system is being shut down at the end of its operational period, the last module executed should again give the systems administrator exclusive control to gather any data or set/inhibit/enable any indicators necessary for the next day's operation.

- E. In the event the application terminates during the day, a special module must clean up any outstanding messages, close files, and record the accounting statistics before relinquishing control back to the operating system. In addition, this module should probably communicate with the system administrator for any last minute input prior to relinquishing control.

- F. If the system is reconfigured during the day, new configuration tables should be written on the log and any accumulating measurements should be reinitialized.

108. Usage and Service

For many years the administrators of large host systems have configured to their average workload in an attempt to keep the systems fully occupied 24 hours a day. They have been encouraged to operate in this manner by auditors who were acutely aware of the cost of a central computer facility.

In contrast, on-line systems are traditionally configured for their peak loads. This guarantees that terminal response time will not suffer during periods of maximum activity, and as a corollary leaves some unused capacity during their off-peak hours. Some designers resist attempts by financial managers to use this surplus capacity as it always results in queues, priorities, and complexity. Other designers attempt to manipulate the real world so the peaks are not so pronounced and the peak hour tends to approximate the average load. Sometimes a compromise is in order if configuring for the peak requires hardware or software that may not be available, or if the price difference between a configuration for the peak and an average configuration is drastically different.

If an attempt has been made to reduce the peak load, but if the remaining load still taxes the installed capacity, the designer can build a transaction dispatcher on the front of his application that appends a priority to each transaction upon receipt and queues it appropriately. Then the queues can be processed in accordance with the available capacity, and the designer need not depend on special conditioned behavior by the workforce.

109. Human Factors

A good on-line system design couples the computer intimately to the workforce at the terminals. There are several systems aspects which promote that coupling. The following considerations are typical:

A. Some operator commands imply a secondary action; i.e., "if the number of persons taking a tour is changed, the price of the tour package must be recalculated." The application's command processor should be programmed so the secondary action is automatic unless cancelled by the operator. To do otherwise would allow an untrained or inattentive operator to affect the integrity of the files.

B. Similarly, if the system is carrying on an interactive dialog with an operator and if the next information request is obvious, the computer should be programmed to present the next message to the operator voluntarily without waiting for operator input.

C. Many systems contain a table of authorized users and their access privileges. Thus at sign-on time, the operator's name is located in the table before access is given and if the name matches, the privileges allowed this operator are authorized.

More sophisticated systems add a code to this table which describes the operator's level of proficiency. Operator prompts, dialog, and error comments are then prepared assuming two or more proficiency levels. Then whenever it is necessary to communicate with the operator, the error text is chosen which contains the abbreviations/full text messages appropriate to the operator's degree of proficiency.

Including the programs to set and change these proficiency codes, such a system requires more programming than do systems with a single level of proficiency. However, for high volume systems involving many highly proficient personnel; for large systems involving constant turnover and hence a continuing portion of inexperienced persons among the workforce; and for situations where supervisors, managers, and executives require occasional infrequent direct access to the system but never develop the proficiency to decode abbreviated messages; it is probably worthwhile to add extra programming to cater to widely diverse groups.

D. Along the same line, extensive coding or heavy abbreviation in computer responses usually implies a need for increased operator training and a code book to decode the abbreviations. Some designers code the responses heavily for the proficient operator, and then implement a MORE command which causes the full text of the appropriate code book section to be printed out on-line whenever requested.

While such programming may be necessary for infrequently occurring error cases, the savings in character transmission time and the savings in characters displaced on the screens are so small that the cost of the extra programming is never recovered. Alternatively if the message back to the operator is carefully edited so the subject appears in the first few words, the experienced operator can react to these first words, while the inexperienced operator will have the full text to read.

E. The administrative system surrounding the computer should be designed concurrently with the computer programs. If paper files are kept or human assistance is required to complete some function, a general rule dictates that the computer be programmed to make the human's job easier. This is true even when extensive computer programming is required. An example will illustrate this point: If the computer files are not in sequence on the same key that is used for references in the external physical world, sort the outputs from computer processing into real world sequence before printing or queueing for display.

F. If one of the primary purposes of an on-line system is to support on-line query, the designer should seriously consider allowing each authorized user to catalog search commands for repeated use. Thus a stock broker monitoring the performance of the market for a series of clients or the flight dispatcher monitoring the status of all departing planes, or a physician inquiring about certain patients, could enter queries into their private catalog. Some personal shorthand notation could then be used to invoke those queries and receive the computer's response.

Obviously the ability to compose a query on-line and have it immediately executed is basic and must be programmed first. But the addition of a SAVE command and a mechanism to create, modify, and invoke commands from a personal library will save keystrokes, reduce the in-bound communications load, and reduce the errors over a system that requires the rekeying of frequently used command strings.

110. Administrative Features

A lean, unadorned, strictly functional network is hard to administer because the persons responsible for its administration cannot observe what is happening and may not even know all the principal players. Designers concerned about system administration should consider the following:

A. Establish one cell in each node to count errors. Then as each operator message is constructed, increment the cell and append it to the message. The range of numbers assigned to error messages in the course of an hour, a shift, or a day is a gross measure of trouble. Further, the message numbers themselves provide a unique ID for a manually kept trouble log.

B. If an application is designed so that all the message traffic between the application and the terminal operators goes through a single module, the stage is set for the construction of a useful debug tool. The module can log all traffic through it and append codes to each message log entry indicating which party (the computer or the person) originated the message, and which terminal-operator pair is involved in the dialog. The message logs can be selectively printed out by terminal and the result will read like a verbatim transcript of a conversation. The monitor module should be programmed so it can optionally log the dialog, print the dialog dynamically if a hard copy printer is available, or display the dialog to the system administrator.

The system administrator should be able to request that the monitor ignore all the traffic and make no logs, select only the messages involving a specific person, select only the messages involving a single terminal, select all the messages involving a single loop or communications line, or record all of the system's interactive dialog regardless of source.

During debugging the monitor module can be set to capture information for debugging purposes, and after the system is in operation, the module can be reenabled any time a terminal or line proves troublesome. If the system security requirements are tight, the security administrator can use the module to intercept unauthorized traffic. If the monitor module is left in place after the code is debugged and enters production, it will be available for continuing troubleshooting should the system encounter difficulties in the future.

C. If several processors are interconnected in a net, the computer console operators will benefit from a high priority, operator-to-operator, hard-copy communications mode. Thus operators can pass details around the network and a hard-copy audit record is automatically created. To produce a complete audit trail and to guarantee its integrity so off-shift problems can be diagnosed the following day, each message should carry sender, receiver, date and time, and be sequentially numbered.

D. Small systems are frequently installed with the minimum number of necessary hardware devices. If the only printer attached to a node goes down, the entire site can be impacted if clerical procedures are vitally dependent upon hard-copy output. Therefore a good practice would call for a header on each print file containing the site identification, the report name, the number of pages in the report, the preprinted form number this report expects, the program producing the report, the date the report was produced, the security classification of the data in the report, and any instructions for special handling. This header should be printed as the first page of the report, since it uniquely identifies the report and provides instructions to the operator concerning the handling of the printed copy. Furthermore, if it was necessary to transmit the entire report to an adjacent node for remote printing, the report would be self-defining and would likely be handled properly at the remote printer and properly returned to the originating node.

E. If data is distributed throughout a network, the replicated data bases must sometimes be synchronized. If the system cannot be shut down but if update transactions can be temporarily deferred, provide for a queue at each local node to capture update transactions against a data base that is in HOLD status. Further, program the inquiry processes to respect the HOLD status and to display a comment as part of an interactive query which says: "Data current to _____, may not contain latest information."

111. Applications Pass-through

Some vendors offer software containing an emulation feature which allows a remote system and its terminals to replace a hard-wired control unit and its terminals. While this feature was designed to assist in migrating to a distributed environment, the principle involved is useful for other purposes. If the data files supporting an inquiry application reside at two or more locations in a network, a small applications program can be written which will allow the terminals attached to one node to inquire of the files attached to an adjacent node if the files at the first node are out of service. While users may experience some performance degradation, the applications pass-through feature will allow critical work to be processed.

This same feature could also be used whenever extra recording or logging of terminal operator dialogs was desired, and the node to which the terminals were attached was configured without the necessary devices to support the required logging.

If such pass-through code were programmed so it selectively routed messages for local or remote processing, depending on the address of the device where the message originated, it would then be a valuable tool to support the installation of new hardware and software. The routing could be set so all production transactions were processed remotely while one or two local terminals tested for correct installation of the latest version of the software or the latest modification to the data base.

112. Network Administration

If contemplating the design and installation of a large multi-node network, the design team must address the problems of network administration early in the design cycle or they will be unconsciously foreclosing some of their options. For example, consider the following sample topics:

A. If contemplating central management of a network, give careful attention to the on-site administrator's manual at

128 Design Phase

each node which covers startup, shutdown, restart, housekeeping, first level problem diagnosis, simple repair, problem reporting, and fix testing.

Also consider the central site administrator's manual which must be complementary in every aspect to the remote site manuals.

Then proceed to design so the proper information is presented to the proper party at the appropriate time so all operational conditions are covered.

B. A central network support center must have a library which contains multiple data sets. Tools and procedures must be in place and personnel trained so the various members of this library can be dynamically maintained. As a minimum the library should contain data sets for production programs, processes, trouble symptoms, and problem determination protocols.

C. In designing a network, provision should be made for broadcast messages. The system administrator should be the only one authorized to create broadcast messages and he should use this feature sparingly. A well designed system would provide two priorities for broadcast messages: immediate (e.g., interrupt the session at the terminal to present the message); and as a second priority, a message to be presented at the next breakpoint in the process for individual users (for example, sign-on, sign-off, transaction complete, request a new data base, etc.).

The broadcast messages provided by the system administrator should be sequentially numbered and stamped with the time and date of origination. The system administrator should have an option which allows him to request a positive acknowledgment from all users logged on the system. Some provision must also be made for maintaining a file of broadcast messages so users not on the system during the period when the message was active can request (by broadcast message number) any they may have missed.

D. The network support center should be able to declare that any device on any node is available for use, is off-line, is down for maintenance, or is assigned exclusively to the support center for test and diagnostic purposes. The support center can then seize a device that is suspected of malfunctioning and if a library of exercise programs is available centrally, transmit the exercise program downline, have it executed, and have the results of that exercise transmitted back to the support center.

Such exercises are not really hardware diagnostics since they do not pinpoint the trouble. They merely transmit test patterns to or from the device and cause programs to be executed. After the results are received at the support center, the center can cause those results to be compared with the expected results to determine if the device, its communications path, and the device-related software are all serviceable.

E. If a vendor provides software for remote service, most of the problem determination aids and most of the hardware diagnostics can be invoked remotely from a central network support center. However, if the teleprocessing connection is malfunctioning, or if the remote processor itself is the cause of the trouble, on-site service by vendor personnel will be required.

If the application programs that run in the local node are not prepared with this on-site service in mind, there is a risk that vital information will be destroyed should the teleprocessing connection require service. On the other hand, if programs, files, and data not vital to the ongoing operation are placed in those areas of storage about to be usurped by the teleprocessing diagnostics, remote problem determination and production operations can be normally restored following on-site service. Otherwise a cold start may be required.

F. The central service center will receive a variety of messages. The message headers must contain information so the messages can be classified upon receipt. The following categories come immediately to mind:
  1. Activity information to be written on a passive log and then optionally processed by an off-line reporting program.
  2. Information from a node, destined for an active program in the support center.

3. Unanticipated information from a node requiring a program in the support center to be activated.

4. Unanticipated information from a node requiring human intervention.

G. Undoubtedly the support center will be programmed to maintain an activity log of the remote events which have been reported to it. This activity log is likely to be hard to interpret due to queues and priorities in the originating node, in any intermediate nodes, and in the support center itself. Thus the sequence of messages placed in the support center log may be different from the sequence of the corresponding events in the field. Therefore each message transmitted should carry the time and date of occurrence, and each message logged should carry the time and date of receipt at the central support center. Then the original sequence of activities can be historically recreated.

113. Communications Administration

The communications problems in a large network require special consideration. Usually message logs, event logs, and error logs are maintained by each node in some combination and by the central support center. These logs are usually sequential and each entry carries a time stamp. However, before the various entries can be reconciled, the communications administrator must know the offsets between the different time-of-day clocks in the network. Further, he needs current configuration information so he can translate terminal and device addresses into communication path information.

If one or more nodes has a communications patch panel to enhance availability, the communications administrator will need access to the reconfiguration logs kept at the site or he will be unable to relate the physical line to the machine address given on the log.

Given all this information, the administrator should be able to isolate troublesome lines, modems, or adaptors, prepare service requests for the telephone companies involved, and set up remote diagnostics which will exercise the respective circuits.

If the network has security features installed, the communication administrator's job is even more difficult due to the protection he must give to the messages that have been logged, to the need to relax access controls so the circuit can be exercised, and if the communications link is encrypted, to the need to bypass the encryption devices during diagnosis.

114. Terminal Response

To establish an intimate dialog between the man and the machine, human factors must be considered early in the design. In many on-line systems the user is completely buffered from the CPU (except in the case of a complete system failure), and the computer's response always seems to be instantaneous. In other systems the computer is more heavily loaded (or some of the transactions involve a significant amount of processing) so the computer's response cannot always be instantaneous. The designer must estimate response time for every transaction very early in the design process as it is a fundamental mistake to treat these two cases in the same manner.

In the latter case the designer must estimate the best case and worst case response times for every transaction. If the best case and the worst case response times are equal, the designer can proceed to the next phase of the analysis. However, if the worst case is materially different from the best case, the designer must determine how frequently this occurs. If the conditions are such that the system saturates on a daily basis, and if the response time goes from best case condition at 8:30 in the morning to worst case condition at 2:00 in the afternoon, a severe problem is at hand.

There are only two acceptable ways for handling this problem (other than more hardware or removing some load from the system). If the operator can live with worst case performance at all times, then one acceptable solution is to normalize the response time to worst case by padding the best case response with some work unrelated to this transaction, or inserting a delay loop if there is no other second priority work available to be done. In either event the user sees consistent response throughout the day, the response is adequate for the intended purpose, and the response is independent of load.

The other acceptable technique is to give the operator intermediate progress reports. If the worst case differs from the best case by a few dozen seconds, the computer can be programmed to provide a blinking indicator on the screen so the operator knows work is being successfully performed in his behalf. Another solution is to estimate the delay and program a clock display on the screen which counts up as long as the computer is making progress in the computation.

One large data base system instantly handled queries where the file key was known and provided an instant estimate of response time in the event a file search was required. As the file search took place, the system provided the operator with service messages which indicated when 25% of the file had been searched, when 50% had been searched, etc.

Another on-line system was using hard-copy terminals with the computer programmed to shift the terminal from upper case to lower case periodically as an indication of progress. Thus the operator got an audible physical signal that progress was being made until the computer could provide the requested response.

The worst human factors situation occurs when the computer provides no indication of progress and the operators become concerned because the response time varies. In these cases the operators press the Enter key repeatedly on a mistaken assumption that the last message may not have been transmitted to the computer. This caused a modification in at least one on-line car rental system, since the computer had to be instructed that the depression of an Enter key when no data was transmitted merely meant the operator was impatient. (Interestingly this slows down the very process the operator is trying to speed up.)

The system designer, therefore, needs to understand the "mental set" of the operator so the computer can either live up to the operator's expectations or provide the operator with appropriate feedback in the event those expectations cannot be met.

115. Estimating Response

After the transactions have been enumerated, the work corresponding to each transaction has been detailed, and the terminal response has been formulated, the systems designer can estimate turnaround time for each transaction.

If the requirements analysis was conducted as suggested by Chapter 3, the designer will know approximately how many transactions of each type can be expected during the day, shift, and peak hour.

The previous section suggested that a user's manual be drafted shortly after the preliminary design was completed. If this was done, the design team then has the three pieces necessary to construct processing scenarios to cover:

- morning startup
- routine processing
- evening cleanup
- abnormal cyclic processing
- special runs
- recovery actions

Each scenario should describe the man-machine dialog that takes place to complete each transaction and present a sample of a person's input and the computer's response. If the margins of the worksheet are annotated with time delays, the person's productivity can be estimated, and any critical response time situations can be identified.

Naturally for small systems or systems with a low input volume, or inquiry-only systems where any reasonable response is satisfactory, the design analysis outlined above is optional. However, for large distributed systems involving many nodes with several terminals on each, facing a variety of workloads and environments, the response time scenarios will allow human factors problems to be discovered before the system is coded, tested, and found unacceptable by field personnel.

116. Dialog Designing

Many factors enter into the design of a man-machine dialog. The most important of these is the designer's ability to picture himself in the role of the terminal user trying to enter, query, or change the data contained in the machine. Several design hints have been distilled from past experience:

A. When entering data on screens, correcting, or displaying inquiries, stay with consistent formats throughout to ease operator training and skills required. Also be consistent with button sequences even if slightly more keystrokes are required. The consistency will more than make up for the extra strokes.

B. Be consistent in handling options on CRT input screens. Phrase questions so a yes answer always implies the same action, and provide consistent methods for backing up a record or a screen, for skipping a field, for skipping to the end of a screen, for recalling a previous menu, etc. Set up these basic protocols first and then do the applications screens within this framework.

C. On a fill-in-blanks type screen, display the whole screen at time of selection. Do not build the screen as blanks are completed.

D. Human factors are better served if display screens are updated incrementally, sweeping from top to bottom, rather than in burst mode where the entire screen is simultaneously refreshed.

E. If long search processes are allowed, they should be interruptible. At the point of interrupt, the operator should have the option to request the status of the search and receive the context of the search (i.e., search request, files involved, currency of the data, etc.), and a statistical status report on the search in progress, i.e., statistical summary of the results that have been presented, and statistics on (or some other method of estimating) the search effort which remains to be accomplished. Alternatively, he should have the option of discontinuing the search to proceed on a new tack; or he should have the option to direct the entire set of search results (not just those that remain) to a print queue for delayed printing; or, of course, the option to resume the search at the point of interruption.

117. Human Factors Hints

Experience also yields some hints to be used in the design of the man-machine dialog itself:

A. Keep date and time of last update in the data base. Display this information on screens and reports to identify currency of information presented.

B. When designing interactive commercial systems, segregate all transactions into uniform classes based on complexity and the terminal operator's tolerance for delay. Design the system internally to provide uniform response to each transaction class regardless of load. Keep a dynamic measure of capacity by class. If out of capacity, dynamically prohibit further terminal log-ons until the load decreases.

C. Operating systems software should provide service priorities. Applications should be designed to place production keyboarding at the highest priority, with other man-machine dialogs next, telecommunications next to that, and electromechanical devices (such as printers) last. Then deterioration in print performance is an indication that saturation is occurring.

   If saturation occurs, system tuning will usually result in temporary relief until the workload outgrows the installed configuration.

   If the production keyboarding (data capture) is not given the highest priority available, then other activities can interfere with the keyboard rhythm. This will doubly impact productivity, first by directly slowing down the input process, and second by causing operator errors since the rhythm of the keyboard is broken.

D. When inputting on a screen, request the data in the units most natural to the user, allowing the computer to perform the necessary conversion. Always display the units alongside the input variables.

E. For fill-in-blanks record creation, split the screen and set aside a small area for machine-man dialog. This will allow the machine to initiate (or the operator to request) information messages without disturbing a partially completed screen.

F. Choose a consistent vocabulary and use it uniformly for operator prompts and CRT-to-operator instructions. Be sure to use special punctuation sparingly and consistently on

CRT screens. Words like 'Avenue' and 'Ave.', or 'January' and 'Jan.' should be linked and treated as synonyms.

G. Punctuate or otherwise break long fields of numbers to improve readability and enhance the human to human communication: 213/871-4320 or 530-74-0447.

H. Establish a convention for handling missing data: .00 for dollar amounts, ... for alphabetic characters, - - for an alphabetic field.

I. When the computer displays a variable length list on CRT, use a standard technique to indicate end-of-list.

J. When an error is detected in a small transaction, it is usually acceptable to reject the entire transaction and have it reentered correctly. However, if the transaction is large or complex, allow rekeying of just the fields in error. If a single field is sufficiently large and complex, provide partial field correction.

K. For applications with multiple CRT screens, adopt good reports control procedures and put a unique ID on each screen for reference purposes.

L. Consider adopting the following good protocol: any record to be modified must first be viewed.

M. Sometimes CRT screens are used for data entry where the paper input data forms are not subject to change. If the input data forms cannot be changed, it is preferable to lay out the screen to correspond to the input data form, rather than asking an operator to extract fields randomly from the input data form for entry into some "optimum" screen design.

N. In prompted interactive dialogs, provide a fast path for highly skilled terminal operators.

O. Build a display option into the end of message processing in a data capture application. Then if the operator becomes confused or distracted, he need merely request a display of the last transaction entered to regain his place in the sequence.

When the transaction is displayed, the keyboard should be enabled for correction or end of message so the operator may verify the transaction entered and either change it or direct the system to accept it.

P. When presenting search results on CRT screens, label every variable and give its units.

Q. In a big system give attention to the human factors of error/diagnostic message displays to avoid saturating the console operator with uncorrelated data. Also, since the operator may have been distracted, or since a replacement operator may have just reported for duty, be sure to present the full context (system ID, application name, module recognizing error, data sets open, user input or request that caused the error, etc.) so the message can be analyzed in context.

118. Reports Design

An experienced data processing person can deal with reports in any format and style (even hexadecimal dumps) provided they contain the proper information. However, the efficiency of even the experienced person suffers if the report is not well laid out for its intended purpose. In contrast, people without extensive DP training are confused and sometimes misled by reports not exactly tuned to their needs. Some hints for good reports design follow:

A. Every report should have an order. If manual files or processes are involved, the report should be sorted to match the order of the manual process, i.e., if manual pending files are in last name alphabetic sequence so telephone status inquiries can be accommodated, sort computer status reports to last name alphabetic sequence.

B. Good reports design dictates putting the title, unique report ID number, date, and page number on each sheet. The last page should carry item counts and control totals to indicate that the report is complete.

C. When printing lists that contain control breaks, provide extra space following the break to facilitate scanning. As a general rule, each break should provide both item counts and dollar totals. Grand totals should be printed at the end of all complex reports.

D. On printed reports, the leftmost column of data should be the primary sort key; next column, secondary key; etc. The sequence of data fields presented in the rightmost columns is a matter of user preference.

E. Unless the decoding of coded variables makes a formal report cluttered and hard to read or causes many pages to be used in lieu of one concise page, decode all coded variables so a casual reader can understand the report without a codebook. When decoded, print both abbreviated meaning and code, e.g., N. Jersey-11 (where 11 is code held in data base). This subliminally presents the codes for shorthand during conversation, for possible use on coded input forms (where the savings may be significant), and in preparation for the occasional informal (coded) report one may see.

F. When designing a system which has extended search capability, it is traditional to allow the terminal operator a quick look at the data and then optionally to request that the entire results of the search be printed on the nearest printer. If that option is provided, be sure to print a header sheet on each set of search output which gives the searcher's name, his mailing address, the date and time of search, the search request itself, and the date and time of the last update to the data base. After the entire set of responses is printed, print an item count or a notation such as "End of Listing" to identify the last page of the search report.

119. Part-time Console Operators

In a network, the processors in each node may operate unattended for long periods of time. Some systems may even be designed with operators on call to the system when intervention is required. In other instances the system may be attended by a person without depth of experience in data processing. Not only will documentation have to be carefully prepared so it can be understood by casual operators, but some additional features will need to be supplied so the operator can quickly get back into context whenever he returns to the machine. The following list contains some examples of features to help the casual operator:

A. The operator must receive information and monitor the following:
  1. Tape log error and status messages.
  2. Data base status.
  3. All error messages.
  4. Any adaptive communication line transmission rate changes.
  5. Security violations.
  6. Lost messages.
  7. Device-not-ready status.
  8. Output from any trace program.
  9. Automatic configuration changes.
  10. Queues exceeding threshold values.

B. The operator will require the following commands:
  1. Reassign printer and card reader.
  2. Sign himself on and off.
  3. Set/reset test status (devices, programs, and system).
  4. Send/receive test messages.
  5. Display status (system, test, maintenance, or available).
  6. Restart node and set date and time.
  7. Graceful shutdown.
  8. Display dump on request (various canned formats).
  9. Communication with support center or operators at adjacent nodes.
  10. Enter message for broadcast at local node.
  11. Start/stop trace (on a device address, on a loop or line, or on the internals of a program).
  12. Enter values for all control (threshold) variables and configuration assignments.
  13. Start/stop/quiesce/restart any component.
  14. Initiate down line loading from central.
  15. Run special diagnostics on command from central.
  16. Force end of (tape) file and initiate tape change.
  17. Set and clear debugging traps.
  18. Start/stop data base load/recovery.
  19. Start/stop performance measurement.

C. For casual operator surveillance over tape mounting/data base maintenance, the system should provide a short pushdown list of abbreviated tape labels for the last few tapes hung with the top entry being ID of the tape currently mounted.

120. Casual Terminal Operation

Distributed computing allows a company to change the way its data processing services are organized. One of those changes frequently involves placing data entry and inquiry terminals in the hands of the ultimate user while simultaneously placing the responsibility for all those inputs with user management. To gain the promised benefits from this mode of operation, terminal services must be programmed to be "friendly and helpful" so users unskilled in data processing will not be intimidated by the computer.

Previous items have suggested that generic terminal formats be established, that abbreviations be allowed (but not required) on input and be avoided on output, and that a context header be displayed on all multi-screen presentations. Additional thoughts in that same vein are:

A. If the system has a table which controls the award of privileges to each authorized operator, the command VIEW VERBS as part of the sign-on sequence would allow an operator to determine if the commands he would be allowed to execute were sufficient for the work to be attempted.

Similarly, the command DESCRIBE <u>verb name</u> would allow an operator to get a current description for any command he was allowed to use, but could not remember the details of its usage. This request, of course, should be available to the operator at any time during his session.

If an abbreviated form of the user's manual was maintained on-line to be available to the DESCRIBE request, the mechanism would be in place to allow an operator to request DESCRIBE <u>error-code.</u> The system could then present coded or abbreviated error descriptions whenever an error was encountered, and these would be suitable for trained operators who use the system frequently. However, the untrained operator, the manager who used the system infrequently, or the trained operator just back from leave or extended vacation could request a narrative description of the error condition and an enumeration of the recovery alternatives.

B. One advanced system asks the operator to state his intentions as part of the log-on sequence. The log-on is then aborted if the operator is not authorized to perform his desired action; e.g., if a trainee wished to log on to a file and edit some data in the production mode, the log-on would be terminated because trainees are not allowed to change production data elements. Similarly, if an employee wished to log on with the special privileges associated only with supervision, he would never get on the system.

C. Some systems carefully check keys and control fields against the master file as part of the initial edit process. If an operator later wished to change an edited field, he would find these identifiers protected so they cannot be routinely modified once entered and edited. Thus the operator could display the record, change the descriptive data fields, or delete the entire record, but could not change the primary identifiers as this would introduce an inconsistency into the data base.

   On the other hand, a supervisor who enjoyed additional privileges could change the primary identifiers. This would cause the record to be deleted from the data base and reentered in the input stream so it could be once again edited in the presence of the master file.

121. Screen Design

After contact with several designers who have concerned themselves with improving the man-machine dialogs, several design hints have been uncovered. Some of these techniques have been presented elsewhere in this handbook. Some additional ones follow:

A. The human factors of a dialog between a human and a CRT screen warrant careful consideration by the systems designer. Before any screens are designed, the designer should prepare an analysis matrix by enumerating the names of all the unique screens to be designed. Then take a large sheet of columnar paper and place the titles of the screens down the left column and the names of the data fields to be displayed across the top. An X at an intersection of a row and column indicates that that data field will appear on the screen corresponding to the row.

After this entire worksheet has been completed, interchange columns of data fields so the frequently referenced data fields move left to lower numbered columns and the infrequently used data fields move right to the higher numbered columns.

Now inspect the matrix and see if a group of fields appears repeatedly on several screens. If so, design a layout and grouping for just these data elements so the presentations will be consistent from screen to screen. Then look for other groupings which, while they may not appear on every screen, appear frequently enough to warrant subgroupings. Find the largest subgroupings and design a portion of a screen format which will accommodate that subgrouping. The remainder of the data elements are isolated unique elements which can be displayed in a third location on the screen.

A section of the screen (usually the bottom few lines) should be allocated to man-machine dialog so the machine can offer questions, hints, advice, or broadcast messages without disturbing the presentation.

Thus in the simple case, the gross screen layout would have an area at the top of the screen for the permanent grouping of data elements (usually the key fields and other identifiers), an area in the middle of the screen for frequently occurring subgroups, an area just below the middle for isolated single data elements, and a few lines at the bottom of the screen for the man-machine dialog.

This is *one* way to lay out screen formats. This procedure corresponds to the preferred procedure for laying out forms. To proceed further, design a screen layout form which has some space for identifying information at the top, has an area which corresponds to the screen raster in the middle (same number of characters and same number of lines), and has some space for comments at the bottom. Now make as many copies of this blank layout as you have screens to design. Using the principles above, lay out the screens on these worksheets. You will have then completed a set of screen designs corresponding to the rules of "positional" layout, i.e., the position of an item on the screen is unique and corresponds to the item or group. Positional layouts are very good if the data elements are fixed length or if the variability in length does not cause lines to overflow.

In a more complex case the simple sum of the maximum lengths of all the variable fields exceeds the screen format, yet the worst case combination fits within the space available. Under this case the preliminary positional layout of a screen must be modified so that the beginning of each area on the screen is allowed to float, depending on the space available after the variable length fields in some predecessor area have been accommodated.

B. To ensure that comments to operators are not confusing, install a table to hold errors and exceptional operator comments on disk and call for them by number. Prior to operator training, edit this comments table as a set for consistency. Further, the table can be easily modified without changing the processing module if it is later found that some comments prove ambiguous.

C. One company frequently has consecutive input records which have many identical fields. They have provided a ditto capability so once a record is entered and stored, they blank only the identifying information but leave all variable data intact in the working buffer. Thus while the operator entering the next record is forced to enter new identifying information, he can selectively overlay any (or all) of the remaining information as is appropriate. This avoids rekeying for many fields.

D. When taking input data from a CRT, two modes of operation are very popular. In one case the initiative lies with the computer which prompts the operator by presenting the identification of the next data field to be entered, and a string of dashes showing the length of the field allowed. (If it is a numeric field, decimal points and commas can also be presented in the schematic blank field.) Thus the screen could say:

Pay Amount $_ _ _._ _!

Then as the operator entered the numeric data, it would be formatted into dollars and cents and would replace both the dashes in the field and the terminal punctuation (the exclamation point in the example given).

Alternatively the initiative can lie with the operator. In this case the cursor would be positioned by the computer, and the operator would enter some identifying mnemonic and follow it with a free form data value. The computer would then interpret the mnemonic, find the corresponding format in an internal table, and then interpret the data value entered in accordance with that format. As feedback, the computer would return:

Pay Amount (PA) $123.70

E. If the population of users varies greatly in training or competence, additional programming will be required. When designing a high volume application for professional input personnel, heavy abbreviation and mnemonics can be used since this cuts the volume of keystroking (and since operators would get tired of reading fully qualified names for all data fields anyway). However, some provision needs to be made to assist the keyboarding supervisor who may have once been a highly skilled operator, but now spends most of the day on administrative matters and cannot remember all the mnemonics. The obvious solution to this is a training mode which will help new operators break in on the system and will help supervisors recollect mnemonics they cannot remember. A HELP capability should be coupled to a training mode.

If the screen has a dialog area set aside as the bottom few lines (discussed above), then the computer can be programmed to recognize a HELP request at any time. Then if the operator temporarily forgets what to do next while completing a screen or solving a problem, the cursor can be moved to the dialog area and HELP requested. Since the computer will already have a pointer positioned for the next logical field to be entered, that pointer position can be used to look up explanatory information from a text file and display that descriptive text in the dialog area.

When dealing with computer programmers, almost the same techniques will suffice. The hotshot coder who is using the system all the time will appreciate crisp abbreviations and mnemonics. A sophisticated system might even allow him to declare his own mnemonics based on current context.

As before, if the programmer enters identifying mnemonics, the feedback should give both the full name (usually abbreviated) and the mnemonic he entered. The system should also accept full names so a system designer or programmer who has been away from the machine for a period of time can use the full formal names. In return the system should supply the formal name and the mnemonics, thus refreshing the programmer's memory of the shorthand notation.

F. If a single complex data item (record) requires more space than is available on the screen, the available screen space must be extended in some way. The preferred way to accomplish this is to break the screen up into three separate sections. The first few lines at the very top of the screen comprise the header and identify the data item. The center of the screen is used as a work area and its format changes as the pages of data are scrolled by. The bottom few lines are reserved, as discussed before, for the man-machine dialog.

The header section of a screen contains more than just the identifiers from the data item. The header on the screen should display information which tells the operator how many pages of data there are in this item and identifies the page now being displayed in the center of the screen.

Thus if text were being displayed, two lines at the top of the screen would give the chapter number, the chapter title, and the total number of lines of text in the chapter. If the third "page" of 20 lines were being displayed in the middle of the screen, they might carry line numbers 61 through 80. At the bottom of the display a few lines would be reserved for the man-machine dialog which can be conducted without overlaying either the header or the current page of information.

G. If variable length fields must be handled and if the field variation is great, some designers find it convenient to provide an area near the bottom of the screen where the variable length field will be initially displayed to provide feedback upon entry. Then after the field has been completely entered and has been initially edited, the operator presses a key to cause the field to be accepted. The computer moves the field from the entry area up to its final resting place on the screen.

This allows the screen geometry to be adaptive so no more space is allocated to a given field than it needs in each instance, while at the same time allowing the operator's eyes to focus on a fixed area and visually verify what has just been keyed.

H. In programming a quick-look capability for an inquiry-search-retrieval system, selected data fields known to be meaningful to the searcher are frequently presented in a fixed format on the screen. Then each line of the screen represents an abbreviated form of the record which corresponded to the search command. If the lines on such a summary screen are sequentially numbered, the operator may be given several options: print the whole report, show the next screen of summary information, show the previous screen of summary information, cancel all the output so the search request can be revised, or selectively print items from the summary screen. Since the program, the display, and the operator are all in the same context, the selection of specific records for format and print can be accomplished by causing the operator to enter only the line numbers from the display. Thus the extensive keystroking is avoided that would be required should the system demand that the operator enter the fully qualified key for the records desired.

122. Data Base Architecture

A data bank is a collection of data bases. A data base is a group of related data sets. A data set, synonymous with a file, is an organized collection of records about some subject. Records are composed of data elements, synonymous with fields, that contain values. If each record is complete in itself and describes some person, event, or article and contains a unique key field and some data elements, and if the data set is logically sequenced on that key field, the result is a flat file as was traditionally held on magnetic tape. If several records are related, the file has a structure, one popular one being the hierarchical organization which has root segments and dependent segments. These are direct analogs of the old punched card technique involving master records and detail records.

In a data base consisting of more than one file, the secondary files may be indexes to the primary file. Thus if the primary manufacturing file were maintained in part number sequence, a secondary file (index) might be held which contained a table of work orders versus part numbers. If the part number were known, the desired record could be directly accessed in the primary file, whereas if only the work order were known, the secondary file would be accessed first. Then a list of one or more part numbers would be obtained and the primary file would be accessed for each part number record in turn.

Every data record contains one or more identification fields which, when taken together, constitute a unique key. The remaining fields in the record carry data or control information. Designers familiar with batch processing usually place control information at the front of the file where it is held once and applies to all the records in the file. Designers familiar with on-line applications, and with file structures peculiar to on-line update in particular, have found it necessary to add control information to each data record so a single access will obtain all key data and control information necessary to allow processing.

The need for both fixed and variable length fields introduces a second structure at the record level. Whenever the length of a single field varies significantly, seasoned designers consider the introduction of field formats which allow the current length of the field to be explicitly indicated with each occurrence. If a system were designed to handle text, the shortest sentence would be three characters and the longest sentence provided for might be 200 characters.

Now it would be possible to store all sentences left justified in a 200-character physical record and use the terminal punctuation at the end of the sentence as a delimiter. However, storage would be wasted if the average sentence length was significantly shorter than 200 characters. Thus to conserve storage, one might place an explicit length field in a fixed location at the beginning of a record so this field could contain the exact character count of the sentence that follows.

All these techniques, and many thousands of variations, are implicit whenever files, data sets, or data banks are mentioned in this handbook. Thus the discussion can be focused on the distributed processing aspects of file design without belaboring the details of the exact file organization most appropriate for any specific application.

It is helpful to consider the structure of data being processed as an analog of the corporate structure. At the risk of stating the obvious, it must be noted that the upper levels of business organizations are relatively stable while the farther down one goes in an organization, the more growth and change affect the organizational structure. The definitions and data elements at the top of the structure are fairly stable (the annual report does not change very much). Yet as one proceeds down the levels in the corporate organization structure, one finds the data structure changes more rapidly. In other words, there is less uncertainty in the aggregate than with the individual components of the business.

Following this analogy further, we find that large centralized data processing shops are relatively stable (both by design and happenstance). However, as we migrate data farther out towards the individual components of the business, the more flexibility is important to allow the accommodation of growth, retrenchment, and dramatic changes in direction.

Several authors have observed that data is distributed in several modes:

A. 100% centralized.
B. 100% decentralized, with one logical structure and no redundancy.

C. Popular portions duplicated at several locations for inquiry, while each piece of data is uniquely owned by some location for update purposes.

D. Summary data stored at a central site with detail data maintained at each node.

It is also observed that a data base can be split on activity patterns to get efficiency, on sensitivity to get security, and on data ownership to get control.

Further observe the degree of redundancy is frequently determined by the allowable response time; the communication sophistication, reliability, and bandwidth; and perhaps by some critical value such as typical order quantity versus usual quantity on hand.

It should also be observed that the way data is handled sometimes changes over time. For instance, an airplane reservation for an individual (or even a group) can be treated informally weeks ahead of departure time. However, as the flight gets within a few hours or minutes of boarding, an on-line time-current update is required.

Data cannot be discussed independent of the transaction streams that impinge upon it. If the transactions are simple and are processed to completion, and if the computer system is lightly loaded, with no queues, almost any design will suffice. If online update can be avoided and the transactions are read-only, with data captured for deferred update, the system is simpler still.

However, when one transaction can generate additional transactions and when the results of several transactions must be linked before the final process or final output can be constructed, then it is necessary for several asynchronous processes to be simultaneously driven, outputs to be collected, and results presented. Further, if one of the processing streams is interrupted, it may be necessary to reverse the effect of all the related transactions that may have already been processed. Multiple processes are not to be accepted lightly, but sometimes they are required if a single processing stream cannot meet the response time requirement. Distributed multiple streams are even more complex, but are sometimes required if the data is distributed to multiple nodes.

A further complexity is added if a second transaction from any location can reference the same record as was just processed from the data base. In the case of the on-line reservation system, any location may book a seat on any flight at any time. Therefore, two or more locations may be simultaneously booking seats on the same flight. Obviously in this environment the transactions must run to completion and the files must be updated before the transaction is considered complete. In this case the designer must provide for many-to-one references and queue on specific key values.

However, in other cases a unique item from the file relates to a unique physical entity. Thus if a car ownership record is being changed from one location, it is highly unlikely that some other location will simultaneously want to change the same record. In this case the environment is intrinsically one-to-one.

Once a designer has considered the environment, analyzed the transactions, estimated the processing required, and performed a preliminary structuring of his data, he should write an applications overview document to disclose his architectural decisions and the reasoning behind them. Unfortunately most programmers are not trained this way. Not only does programmer documentation tend to be skimpy, but even when it exists, the overview chapter is usually missing.

123. Data Base Distribution

Some additional ideas have been distilled from experience to continue the discussion on data base distribution:

A. When distributing data, note some of the natural logical breaks that frequently occur:
   1. Format tables for field editing need not reside with the master files.
   2. Look-up data for initial processing can also be distributed (eligibility data for transaction acceptance or rejection can be remotely stored, with only the exceptions referencing the master file).
   3. The active subset of files for routine processing can be split and distributed while exception processing remains on the host where the full data base resides.

Note: Split files are more complex, use more storage space, provide a more responsive system, and allow routine transactions to be handled without a host.

B. When two files contain common elements, the tendency is to combine the files if:
1. The data elements have common definitions.
2. The file updates can be synchronized on records.
3. The combined file can be accessed on a common key.

Note: Also watch for undue peaking of activity or the side effects of a merge which force interactive update on a larger file.

C. If there are multiple indexes to a master file, and an update occurs, consider carefully whether:
1. Any updates must be on-line.
2. Only the master file must be updated on-line.
3. The master and all indexes require simultaneous change to maintain currency.

To analyze this phenomenon, determine the probability that a second transaction (or string of transactions) will hit a record that was previously changed.

D. A single copy of each data record is efficient even with a high update/inquiry ratio if terminal response can be satisfactory.

E. During the design of files, build a transaction frequency diagram and a transaction sequence diagram. (See Figures 4.2 and 4.3.) Evaluate each alternate file design to determine the seek count for each transaction. Choose the design with the minimum seek count unless worst case circumstances violate some absolute time limit. (Be sure to avoid excess indexes with symbolic pointers if performance is a problem.)

In discussing distributed data, many authors concentrate on data redundancy and fail to give adequate consideration to the differences in transaction streams at different locations. A transaction frequency analysis by location will throw considerable light on this problem.

For example, consider a minicomputer installed in a warehouse. The principal transaction stream would be related to introducing items into inventory and filling orders from inventory. In contrast, the data base at a central site in the same network would be used for ordering and planning purposes. Thus one transaction stream consists of single transactions in an interactive dialog and the other consists of batch processing by type of part, supplier, lead time, or some other characteristic.

The systems analyst should spend some time studying the organization and the tasks it performs before attempting to structure the data.

124. Data Base Design

Experience also provides some hints for the data base designer:

A. The designer should determine if the data base design holds space for all data elements required for management reporting, in addition to those for transaction processing. (If you ever intend to add data elements, leave space now, thus avoiding multiple changes and keeping performance figures from being optimistic with smaller than life data bases.)

An example will help: One hotel had a billing system that collected charges and billed credit customers. When they got around to management reporting, they had to add several fields to each record of charge to contain date-time-location of the department originating the charge ticket. Without this information they could not do a profitability analysis on departments (like food and beverage) which served from several locations, and they could not do a labor productivity analysis by employee.

B. If a data base has been partitioned so data is stored nonredundantly at each of several nodes, then the system should automatically append the identification of the node retaining the data record. Thus if the data being displayed is local data, the ID number of the local node will be displayed along with it. If the data came from an adjacent node, the number of that node will be displayed. The system must retain the information internally so updates can be routed. Even if the application uses data in a read-only mode, the identification of the node permanently storing the data should be displayed as an aid in problem determination.

C. Every on-line file should contain the date and time of last update. This information should appear on all reports and should be displayed, either automatically or on request, to terminal operators. When dealing with accounting files or other data which has a fixed closing date, keep two dates in each file, closing date (date of file freeze) and date of last update, as an aid to balancing and reconciliation.

D. Some data base organizations contain multiple secondary indexes which point to the primary master file. Thus if a personnel file was being maintained by employee number, secondary indexes might be constructed to provide a list of employee names and ID numbers when the department number was known, or to provide the reverse index when a telephone number was known and the employee's name and ID number were desired.

If enough secondary indexes are created, a designer may observe that the amount of storage devoted to indexes has become significant when compared to the space occupied by the master file. If disk storage is tight, there is a temptation to remove data elements from the master file that are redundantly stored in the index files. This decision should be carefully considered because a complete record on an individual will no longer be available from a single source if redundant fields are deleted from the master file. A record must be reconstructed through multiple Seek-Read sequences which rebuild the master record (if it is ever required in one place) by combining the remains of the master record and the data from the several indexes.

While placing additional data elements in the index files (so they contain both the link and the data most frequently requested) improves performance, it also complicates the update process, makes the entire master record more expensive to reconstruct whenever it is needed, and complicates the restart process.

On the other hand, if the master records are left intact and fields of redundant information are added to the indexes, file reconstruction is actually made easier since the most important fields of the master record can be obtained from the indexes. Thus if the master record survives the indexes, or the indexes survive the master record, one can be reconstructed from the other. The only reference to backup files would be to obtain the little-used fields required to complete the master record. The adding of redundant data elements to the indexes then makes master file reconstruction easier and reduces the blackout period that follows a catastrophe.

E. The previous item discussed how to partially reconstruct a master record from secondary indexes which contain redundant data fields. If each master record carried a flag, a bit could be set in that flag whenever a record was incomplete. Thus a processing program would be warned that although the frequently referenced data fields (those held in common with the indexes) were available for use, the infrequently referenced fields were not dependable until they had been restored from the previous copy of the master file or from some other source. Given a completeness flag in every record, production could be resumed quickly while the remainder of the file was being rebuilt.

In addition, if a second flag bit and a checksum were added to each record, then the individual records would contain sufficient redundancy to allow their integrity to be checked. Following a catastrophe, an analysis program could be loaded which recomputed the checksums for every master record and set the flag bit to indicate whether the record was intact. Thus a master file which experienced a power failure during a write sequence could be edited. If errors were found, the records involved in the write could have their flag bits set to inhibit further processing until they were reconstructed, while the remainder of the file would be available for immediate access. Record checksums cost computer cycles, but if input/output processing rates limit the throughput of the system, the cycles are available. Using them for checksums increases the system's integrity and reduces the blackout period terminal users will experience.

125. Data Base Performance

Since the interaction between a program and its data frequently determines throughput, performance is a prime consideration in data base design. Consider the following:

A. If a data base has transaction based recovery, watch the more complex transactions that hit multiple data bases or a data base plus one or more indexes. With distributed data these transactions are difficult to handle since they involve lock/unlock, audit trails, checkpoint records, and synchronizing.

B. Consider restart, reconstruct, and security prior to detailed transaction design and record layout. Be sure journals necessary for file rebuild are survivable and do not unduly impact performance.

C. Analyze the sequence of I/O calls; if records are required in sequence and if storage space is available, increase the blocking factor to reduce the number of I/Os. Also, if several transactions are likely to require the same master record, search the output buffer before issuing the redundant read to the master file.

126. File Loading and Recovery

Several years ago a highly regarded independent software house built a file management system which had beautiful retrieval properties on almost any key, but took hours to invert an updated flat file and load the data base. Many modern authors missed that lesson since data base loading is not extensively treated in the literature. However, the data base must be initially loaded when the system is installed, and, since that process seldom goes perfectly and since the application programs are seldom checked out initially, the loading is repeated a few times during the first months the system is in operation.

Further, most small and medium systems still use the dump/restore technique for catastrophe protection. If a minimal configuration were installed at the outlying nodes of a distributed network, and if some nodes lacked a second disk unit or a magnetic tape to support dump and restore, a designer might choose to maintain a duplicate copy of the master file on the host, or on some upstream node which had surplus capacity, by sending all update transactions to the host after they had been reflected in the local copy of the data base. If the host maintained a backup copy of the data base on magnetic tape and queued the update transactions for periodic batch update, then recovery from a catastrophe would require repair of the equipment at the local node, update of the backup copy at the host with the queue of pending transactions, and downline load of all or part of the now current data base to provide for complete or partial file reconstruction at the node.

To minimize the blackout at the node, each record should have some redundancy (checksums) so the records in error can be recognized and the entire file need not be completely reloaded. A few hundred records can easily be sent down a communication line of almost any bandwidth. A complete reload requiring 100,000 records would probably result in an unacceptable blackout.

To continue this discussion, consider the following items:

A. If file loading is to proceed concurrently with other processing, priority in the data base management system must allow loading to take precedence. Controls and status indicators must recognize that files will exist in the following states:
   1. Ready, information current
   2. Ready, information not current
   3. Not ready, loading
   4. Nonexistent

Application software must determine the state of each file at open time and queue on State = 1 unless it is part of the reconstruction/loading process. If data has been queued awaiting a file reload, the queue must be processed to move from State 2 to 1.

When processing an accumulation of on-line transactions in a batch update, all normal response messages that usually go back to the operator must be fielded by the batch update program.

The final step in the load process is to synchronize the log tape with the reloaded file. Then the file can be asserted to be in State 1.

B. If files are to be unloaded in parallel with continuing processing, all transactions which affect records behind the dynamic copy pointer must be queued and appended to the unloaded file or these transactions may be lost in the event of restart.

If date-time or message number is kept in each record, and if the log tape carries these same fields, unload during update can occur if an intelligent update from the log tape is provided. This avoids the double update problem in the event of reload.

127. Recovery Logs

The interplay between the redundancy in the master files and the log of update transactions must be carefully defined if the data base is to be logically restartable within the periods of blackout acceptable to the user. Consider the following:

A. Redundancy is needed in files to allow partial file rebuild. Files also need partial file lock while rebuilding.

B. Every transaction (set) that affects the data base should be logged. Retain the fully qualified key to the master record on the log. Append the operator ID, date, and time to the low order end of the key. To get a history on a master record, sort the chronological log on this augmented key. Then the history can be used to support partial file rebuild.

C. Records on the log file should contain both old and new values for fields being modified. This will allow changes to be backed out or files to be reconstructed by simple field replacement without full scale transaction reprocessing.

128. Data Dictionaries

A dictionary is the key to compatible development of computer programs, which are related by function or shared data, in two different locations. The dictionary procedures should support the following:

A. Data element definitions (original and modified).
B. Data set definition and structure.
C. Date and level control on data definitions.
D. Use of definitions by compilers.
E. Automatic criss-cross lists of data elements vs. user programs, and user programs vs. data elements.
F. Edition/revision numbers on user programs.
G. Interlocks to prohibit access to the data base by programs not compiled using the current data definitions.
H. The ability for the data base administrator to list all cataloged programs with compilation dates earlier than the modification level dates on the data elements they reference.

Note: A properly designed data dictionary could also be used to drive conversion programs, audit programs, and listing programs. For a data dictionary to properly support the above functions, its use must be mandatory, it should be an integral part of the data base management system, and the maintenance of the dictionary entries themselves should be automatic and on-line.

129. File/Communications Tradeoffs

Most networks will be geographically dispersed and internode communications will depend on telephone company circuits. Many applications designers will have had no experience with the telephone system, its analog signalling technology, or with the wide variety of modems that are available to transmit digital information across telephone lines. In addition to these analog services, the phone company offers an all-digital interconnect service between 96 metropolitan areas representing 350 cities.

A designer who has no experience in communications would be well advised to concentrate on problem definition and then seek out a specialist to assist in defining alternatives and choosing the solution. If the designer is employed by a company which has a large host facility, it is likely that a communication specialist can be obtained from the technical services group attached to the computer center. Of course contract help can be found on the outside, or if time is available, communication courses are offered, seminars are always being held, and texts for self-study are obtainable.

When considering the communications and network aspects of a distributed application, the following points will be helpful:

A. Sometimes stating the obvious is a useful exercise. A distributed commercial system provides three generic services:
   1. Message switching
   2. Interactive dialog
   3. Batch processing

   Message switching is loosely coupled to the other two services and has independent files and queues of its own. Interactive and batch modes are tightly coupled through the file structures they support. History indicates that a typical interactive system may still have 60% of its work performed in the batch mode.

   If one file structure cannot support both batch and interactive modes satisfactorily, then two options are available. Two simple structures can be created, one for interactive and one for batch, with update transactions being passed between the two processes so the files stay synchronized. If this strategy is chosen, the interactive files will probably be a subset of the batch files. The other option is to organize an omnibus data base and then add enough indexes so both types of processes can obtain the data they need.

   It should be noted that extra file space is required to support both options. In one case the file space is used for holding the redundant subset of the data to support the interactive dialog; in the other case the file space is used to hold indexes which provide alternate access paths into the data. Note also that the multiply indexed data base also requires additional programming to create and maintain the indexes, check the indexes after a failure, and reconstruct any indexes found to be flawed.

   Finally, it should be noted that a local data base is usually favored for supporting the interactive dialog if response time is a problem. Further, side benefits are derived from a properly designed local data base which will improve the reliability and reduce the blackout time following a system failure.

B. It has been claimed that distributed data bases will save communications dollars. Rather than accepting this claim at face value, the system designer must recognize that popular theories are not always the cheapest. Communications bandwidth comes in fixed increments: 300B, 1200B, 2400B, --- 56KB. Note that these speeds are not a power series. Thus if you require a given amount of bandwidth, you should cost out one or two lines of a slower speed versus the next step up in line speed, electronics, and complexity. If you can avoid one of those steps, you may save communications dollars.

The same phenomenon affects computer equipment. Storage comes in modules, disks come in units, and CPUs come with limited speed and channel capacity. To add the second disk to a one disk system is simple because the manufacturer planned for that kind of growth. To add a disk to a system that is already at its limit is difficult and expensive. To take one fully loaded system and split the processing so it runs on two systems requires design and programming in addition to more hardware and software.

Thus the systems designer should be aware that some thresholds cost more than others, and that packing data on a disk that is 80% allocated may reduce the allocation to 50% (and buy some expansion for the future), but it does not make the disk go away.

C. In a recent paper, a network communications specialist* provided the following checklist as a starting point for network traffic analysis:
1. The geographical location of every node and terminal.
2. The amount of data transmitted to or from each terminal.
3. The required response time at each terminal.

---

*Integration of Advanced Communication Techniques, Roshan Lal Sharma, Rockwell Intl.

4. The growth rates in terms of locations, terminals, and traffic counts.
5. Any restrictions on the locations of multiplexors and concentrators.
6. Any restrictions in the selection of communications tariffs (e.g., AT&T, WU, Datran, MCI, Telenet, etc.).
7. The limitations on communications costs based on any preliminary allocations that may be known.

D. Just before the traffic analysis is completed and the bandwidth analysis is ready to begin, go back and check for exception conditions. Be sure that problem determination from the remote location will not take excessive time due to a narrow bandwidth, or that a narrow bandwidth renders it impossible to refresh a data base after a file failure in a node.

If the transmission times are too long for some of these abnormal activities, consider increasing the complexity and compacting data, or editing the data surviving at the node to find out which portion of the data base must be reloaded. These are preferable alternatives to buying additional communications lines so a full data base restore can be accomplished within the time limit.

E. After a communications protocol is selected, modems are chosen, and line speeds are ordered, prepare a graph of line time versus message length for various line speeds. Then review the impact of multiple sessions and message priority and their effect on overall system performance.

130. Communications Protocols

The latest crop of small computer hardware and software is definitely communications oriented. Almost all recent announcements support some of the more sophisticated link protocols and some new software can support the more sophisticated network protocols such as SNA and X.25. Thus a network of intelligent nodes can be configured without writing extensions to the basic communications software provided by the vendors. Some important features are:

A. Protocols like SNA will support multiple simultaneous sessions over a single communications line. Thus in addition to routine data transmission over a line, the same line can be used for:
   1. Downline program load
   2. Operator to operator communications
   3. Remote callout of statistics by a central support center
   4. Remote problem determination

B. The easiest systems to install are those with new hardware and software supplied by a single vendor. This virtually guarantees that the interfaces are compatible. However, many existing systems need to be upgraded piecemeal. A typical case occurs when the processor and control units are upgraded but the existing terminals are retained. In these cases you can buy a processor which has sophisticated communications software, but you cannot always use the latest protocols because your devices understand only Bi-sync or maybe even Start-Stop.

One or two of the vendors provide migration features which are integrated into the communications software to provide compatibility with the older line disciplines and support the attachment of dumb terminals. These offerings are so new that not much field experience has been accumulated. However, the concept of using migration software to provide back-level compatibility is a worthy one. The cautious analyst faced with old terminals and new processors would be wise to run some tests to determine the level of compatibility actually realizable before committing himself to development schedules involving a combination of old and new equipment.

If you are upgrading a banking or credit card network which has hundreds of terminals using an obsolescent protocol, you are in a very unforgiving environment. Whereas some designers may be able to take the vendor's migration software and use it as a way of life in some more forgiving environment, the prudent designer in an unforgiving environment would be wise to probe deeper. The vendor's diagnostics and service features are undoubtedly geared towards his preferred offering, i.e., the more sophisticated protocol. If you plan to use his compatibility package for heavy production, you may have to augment its diagnostics and improve its service features.

C. Under some of the newer protocols, the transmitting location holds a message until its receipt is acknowledged. If the receiver runs out of buffer space, the message is never acknowledged and hence can be retransmitted after the receiving node makes buffer space available.

D. If mixed character sets are required to support existing terminal equipment, and if messages are to be transmitted through a network from the originator through one or more intermediate nodes to the processor serving the addressee, then reserve space in the message header for a protocol flag which defines the terminal, the protocol to be used in the last link between the processor and the terminal, and the character set used by the terminal.

E. For systems involving extensive communications, the designer should consider reserving a space in the header of each message for a compression flag even though data compression may not be initially implemented. Then as processing equipment gets cheaper and more capable, the compression option can be used to trade processing power for bandwidth.

F. Some communications (transactions) may require multiple discrete messages. The transmitting processor should reserve space in storage or on disk to hold the entire set in the event the transmission of any one message is unsuccessful and the entire sequence must be reprocessed.

 The most critical case occurs when a single transaction at a node generates more than one secondary transaction which must be processed at other locations. If all of these transactions are not successfully transmitted and received, then an incompatible file set can result if the entire set is not recalled. Therefore the set of outbound messages should be retained until all have been transmitted and acknowledged.

G. When transmitting program fixes or downline loading program modules or transmitting any other information totally devoid of redundancy, some form of checking is recommended. If the communications protocol used does not perform this checking automatically, checksums are recommended to guarantee that critical transmissions are sent and received correctly.

131. Message Routing

Message routing in a star network or a ring network is straightforward since there is only one path between any two points. Routing is more complex in a mesh network which provides multiple point-to-point paths. Large systems dedicated to message switching contain very sophisticated routing algorithms so the optimum path can be chosen based on line speed, queue length, and instantaneous line conditions. Designers of networks supporting a single set of applications may implement some of the techniques common to more sophisticated dedicated message switching systems. In that vein, several thoughts are offered:

A. In a multi-node network, the message format should be designed with space provided for source, sink, and routing information. If complete routing information is to be specified, the longest path through the system must be identified and space must be allocated to message buffers so each link in the routing may be stated.

    Alternatively the source and the sink can be provided in the message header and each node can be given the option to select the best available path when the message is ready for transmission. Sometimes the two techniques are combined so space is provided in buffers and each transmitting node appends its code to the previous routing sequence. Thus when a message is finally received by the addressee, it contains the actual routing as a trailing part of the message. This routing information can then be used for traffic analysis or audit purposes.

B. If a sophisticated system is designed so each intermediate node is allowed to choose the "best" path automatically, provisions should be built into the software to defeat this automatic routing if necessary so the message will travel as directed. If this is not possible, then remote diagnostics will be difficult to write.

C. The operators of the system should know the reliability of each node, line, and station. Then the routing of very sensitive information can be directed to the most reliable paths. Security codes, priority messages, and program code changes would benefit from a system which allows automatic routing to be defeated so a preferred path can be specified. Then critical messages can be transmitted in the minimum time with the fewest number of errors.

D. Communications management in a multi-node network that contains alternate path routing will be much more complex than in star and ring network configurations. Each node should be programmed to maintain message counts by type for the messages received and transmitted by the node. Further, error statistics on all lines should be maintained in the node. Then when these statistics are transmitted to a central location and are aggregated, the result will be a picture of the network which defines its current reliability and recent error history.

132. Programming Notes

Hints and ideas have been presented throughout this handbook which have implications for programming. Naturally designers who want certain features implemented in a specific way must prepare programming notes as they design, or the programmers, not having all the information and insight of the designer, may implement the feature in some other fashion. The way an application is programmed originally establishes the flexibility for change and growth that remains after the system is running production. Further, the desirability of some features may not be obvious unless specifically highlighted by the designer.

In keeping with the needs of applications running in a distributed systems environment, the following specific programming techniques are discussed:

A. Happily the world is full of transactions which fit easily within available computing capacity and hence can be processed to completion in one step. However, the opposite circumstance can occur; a single transaction can severely tax the installed capacity.

Sometimes this occurs when the available memory must be allocated to the operating system, the communications support, and to one or more applications processing regions. Under these conditions the region size available for applications processing may not be sufficient to handle the largest program required. Even though size constraints may arise only infrequently, they frequently dominate the design of the application system since the structure of the applications programs must allow for all sizes of transaction processing programs from the smallest to the very largest.

If the computer is lightly loaded so excess processing power is available, large transaction processing programs can be partitioned into an overlay structure. The first processing module performs a portion of the processing and all intermediate results are temporarily held in memory, while a second processing module is obtained from disk for execution. Given enough processing capacity (and a relaxed elapsed-time constraint), processing modules can be chained together so an enormous amount of processing can be accomplished on a very small machine. This is even possible with on-line systems if the processing is structured so intermediate results are developed and fed back to maintain the dialog with the user on the terminal.

If sufficient processing power is not available, then the application must be structured differently. In this latter case, one or more transactions are queued until a minibatch is collected. Then as each processing module is fetched from disk, the entire minibatch is processed by that module and a minibatch of intermediate results is produced. The next processing module takes that minibatch and performs the next portion of the work, and so on, until the entire chain of processes has been completed.

In the first case the transaction is held and the chain of programs is fetched and passed by it. In the second case the overhead and I/O activities are significantly reduced since a batch of transactions is processed for each processing module fetched. In applications processing on-line, the size of the process modules is usually many times larger than the size of a transaction record. Thus the second case provides dual economies. First, the total number of program fetches is reduced in direct proportion to the size of the minibatch; second, the amount of information read (transaction records) is usually quite a bit less than the volume of program information to be fetched

A severe problem occurs when the volume of required transactions processing exceeds the available computer capacity and there is also a severe response time constraint. Under these conditions the application designer has three options. First, he can do as much preprocessing as possible so the amount of processing that remains to be accomplished within the time constraint is truly minimized. Second, he can trade some storage space for performance by placing data in disk files in the most readily used format, avoiding packing, encryption, or sophisticated file structures which require CPU cycles to translate from storage format to processing format. And third, the designer can attempt to split the processing into the bare minimum of that which must be performed to support on-line services, and the remainder of the processing which can be deferred (within reason) and processed from a queue at a more convenient time.

Obviously if the application requires each transaction to be processed to completion so the files are instantaneously maintained in a time-current fashion, then none of the processing can be deferred. Sufficient processing power must be installed to handle the transactions within the time allowed. However, if the naturally occurring sequence of transactions indicates that once a transaction intersects with a specific record, no further transactions will occur against that master file key for a finite time period, then split processing is practical. Obviously in any split processing design, enough of the processing must be accomplished in the first task so meaningful information is available for display on the user's terminal, and so the processing context is known and can be saved in a queue for the delayed processing to be completed.

B. Building on-line programs for operation on remote processors requires some advanced thinking about internal program structure. An unstructured program is hard to test and if trouble occurs after entering production, it is hard to troubleshoot. Consider the following suggestions:

1. All dialog with user terminals should pass through a single module. Then by modifying this module, logging, audit trails, and test tools can be accommodated.

    The single module that monitors all user terminal services should logically be as close to the terminal interface as possible, i.e., immediately adjacent to the entrance and exit from the hardware-specific terminal handler. This will allow program drivers (test tooling) to be written that simulate users at terminals and provide a load test for the system.

2. All dialog with the data base should similarly pass through a single module. Not only does this have the same benefits enumerated above, but it allows the processing program to be insulated from changes in the data base. This module can be replaced with one that handles format translations if that should be necessary at some future time.

3. If all user terminal dialogs pass through a single module and all data base activity similarly passes through a single module, the traffic across these two modules can be tapped to give a full activity log, or can be selectively tapped for monitoring or troubleshooting purposes.

4. If these three program structures are built initially, and if the actual code written is designed so it uses an absolute minimum of cycles when the selection parameters are set to zero, then this test tooling can lay dormant in production programs at a cost of some storage space and only a few cycles for every transaction processed. Then if difficulty ever arises, the selection parameters can be reset and sampling can occur without recompiling or otherwise modifying the production program. This approach nearly satisfies the

analyst's needs for gathering troubleshooting data without disturbing the process being monitored.

5. If all error comments are passed through a single module and all events which modify the environment are passed through that same module (e.g., log-ons, sign-offs, tape mounts, etc.), a log of system level events and system error counts can be produced. In addition, this module can be the one which interprets error codes, selects the appropriate text from a table, and builds the message to the operator. This allows error message generation to be centralized so the error text can be edited for consistency and changed to avoid ambiguity. Further, it allows an interactive transaction stream to be queued and processed in the batch mode since the error handling module can be set to field or reroute error comments if users are no longer at their terminals.

6. A single logging module should be provided for use by all service modules and the application programs themselves. The log module can then queue messages for the log file and thereby maintain the sequence of events for proper recording on the log.

C. Table-driven edit modules provide a trade-off between flexibility for change and performance. If edits are table driven, a single table change will allow overlooked conditions to be added without recompiling and retesting the code.

Some managers install programs with table driven functions, and then if performance becomes a problem, review the frequently executed table driven modules to see how much can be saved by recoding them. This two-step process is not always expensive since the alternative may be hard coded modules based on unstable requirements.

D. When programming for maintenance in a big network, no effort should be spared to design stable calling sequences and major module interfaces. Changes should then be controlled so the functions provided by a module can be enhanced without introducing interface incompatibilities. If it is ever necessary to alter an existing interface, it will probably be necessary to change simultaneously all of the modules in the entire network referencing that interface.

E. Lists of allowable codes and tables used for editing and formatting should be designed to be open-ended so *additions* (not changes) can be easily made without impacting the existing program modules.

F. Frequently data records are constructed from a series of transactions during a man-machine dialog. If your system requires this mode of operation, the last step before sending the record to the data base should be a full context edit to verify the validity of the entire record.

If this context edit module is made a separate, callable subroutine, then it can be used anywhere in the system to verify that the various fields in the record are correct and in harmony. During restart this module could be used to verify the integrity of suspicious records anywhere within the system.

G. Many designers have found that structuring the input data area and keeping all variables and control parameters intact during processing costs storage space, but provides meaningful information for debugging and later troubleshooting.

Work areas can be similarly structured so a snapshot at any time would provide a picture of the input being processed and the results accumulated to date. Sometimes work areas need to be overlaid; when this occurs designers should reuse the work area in an orderly fashion so a snapshot is still meaningful. If two or more mappings of the work area are used within any single process, some designers place an indicator in the data area which is dynamically maintained by the processing code so the troubleshooter knows which data map applies to the current pattern.

H. The production control function in a large data processing shop usually has a separate system for distributing reports. This system maintains lists of names, mailing addresses, and numbers of copies for each production report. In small systems some of this concept should be carried over so this very volatile information is not embedded in programs, but is held in separate files where it can be easily changed.

Sometimes the distribution files are so small that a box of 3 x 5 cards will supply the information needed by the printer operator. In other cases a simple list can be automated, and when this list is entered with report number, the system can generate printing and distribution instructions to the operator.

I. Previous items in this handbook have recommended that reports be formally formatted with titles, control numbers, number of pages, and dates describing the currency of the data. It has been suggested that the ordering of columns on the report be chosen so that the fields are printed in sort sequence from left to right; that the report be sequenced to support the manual processes that must mesh with it; and that tallies, totals, item counts, grand totals or some other indication be provided on the last page. To all that should be added an awareness of how the reports will be used and filed so sufficient margins are provided if the reports are to be punched or bound.

133. External Processing

In discussing procedures outside the automated system, this handbook has previously remarked that output reports should be sequenced to support the manual processes involved. For example, if manual pending files are kept in part number sequence so telephone status inquiries can be accommodated, the computer status reports should be sorted to part number sequence. This suggestion is merely one facet of a larger consideration.

In the general case, the automated portion of the system and the manual procedures must be designed concurrently. Thus the forms used for recording input data will be in harmony with the screen formats. The contents of reports and their sequences will support the nonautomated downstream processes. The inquiries needed will imply the secondary indexes needed in the data base. The inquiry responses required will define the data elements to be placed on a screen, and these in turn will indicate the data fields which must be available from status tables or from the data base.

If the total design is properly prepared, the preliminary draft of the users' manual (written immediately following preliminary design) can cover both manual and automated processes. After the detailed design is complete, the user scenarios will encompass time estimates for both the manual and automated processes. The system designer then will be aware of the entire user activity cycle which consists of automated portions, manual portions, and "think time."

Given this information, change to the manual system and its documentation will be naturally considered when change to the automated system is analyzed. Further, training and troubleshooting will be natural fallouts of this analysis discipline.

Admittedly this process delays the onset of programming, but a better system results and development costs are reduced since less rework will be required to make the system acceptable to its users.

134. Network Administration

If a network is to be centrally administered, the following checklist applies:

A. Central problem log.

B. Central fix log.

C. Temporary fix log.

D. Review and approval over hardware fixes.

E. Review and approval over software fixes.

F. Development standards, fix standards, rigid change control.

G. Established problem determination procedure.

H. Analysis of problems to isolate duplicates, warn of potential troubles, etc.

I. Test cases to verify fix application.

J. Positive feedback of fix application experience.

K. At least one member of central service crew ready to travel on short notice.

In addition, the network administrators must be aware of, and sometimes a party to, changes to the manual procedures immediately adjacent to the automated system. Changes to the manual procedures can alter the sequence of transactions in the system, the frequency of occurrence of some types of transactions, and depending on error handling procedures, the gross load on the system.

Similarly, system administrators must acquaint user supervision with their plans for network addition, modification, and enhancement. While users will seldom stubbornly resist an improvement, they may have quite strong feelings on the timing of its installation. In systems that support financial activities, anything that impacts end-of-period closing is frowned upon. In manufacturing systems, adding improvements while new product information is being introduced into the system is discouraged. In general, too many simultaneous changes add undue risk. Some old managers embrace the maxim: "Change only one thing at a time."

135. Global Process Controls

Controls in an automated system frequently suffer from inattention. The users and designers concentrate so much on the function they desire, and on the human factors aspects of the man-machine dialog, that they neglect to consider controls until the system is almost designed or in some cases until the system is installed. Controls are so pervasive that they should be considered early in the design cycle and a consistent control philosophy should be applied.

Most paperwork systems have controls in the form of prenumbered forms, item counts, and batch totals. Where these systems interface with even larger systems, controls take the form of work order numbers, purchase order numbers, and in some cases formal contracts.

When a computer is introduced in the middle of an information system, the flow is interrupted and sometimes the controls are broken. Paper no longer flows through the system from end to end, accumulating signatures and approvals, but stops where the information enters the computer and is usually diverted to a series of chronological or archival files. If the paper flow is random, some designers call for manual procedures which apply a sequential number to each document, allowing the documents to be stored in sequence on that number. The same number is then keyboarded and retained in the computer system as a link. Should the document ever be required, the computer produces the call number so the paper archive files can be accessed. This saves a paper sort in the manual system at a trivial increase in computer processing. This is but one example of built-in system controls. Other suggestions follow:

A. If a moderate-size processor is plagued with severe workload peaking, manual processes can sometimes be scheduled to batch or prioritize transactions so the peak is reduced. To monitor such manual procedures requires the computer to produce transactions counts by time of day so the deferred transactions can be controlled.

B. Where it is possible for two users to simultaneously request access to the same data record, controls must be built in so the first user can obtain exclusive control over the record and prohibit the second user from accessing it, or so the first user is given exclusive authority to change while both users may view the record, or so the second user is informed of the first user's interest at the time he makes his data request.

If some form of exclusive record level control is contemplated, an interesting exception occurs when a user declares himself to be in the training mode. This should modify the exclusive control and allow two users appropriate access to the same record, provided one is a trainee and the other is a supervisor.

C. The literature is fairly complete in its discussions of computer security. Physical security, personnel procedures, access controls, and authorization schemes are rather thoroughly documented. However, one aspect of security that has not been adequately treated is the problem of controls and accountability. These are vastly different in batch systems and in on-line systems. While the batch systems have their problems, most of the problems are known and documented. In contrast, the on-line systems are still so new that distributed accountability, audit trails, and the problems dealing with file quality and certification-for-release are not well documented in the literature.

A previous item on the internal structure of programs suggested that data base access and terminal access go through control modules, and that a common log module be established. If these suggestions are followed, these three key modules can be enhanced to provide almost any level of control or audit trail desired.

136. Specific Process Controls

In addition to the global controls discussed in the previous item, some specifics deserve consideration:

A. If controls are planned from the start, the designer will consider:
   1. Count of transactions from each screen.
   2. Counts on populations in queues.
   3. Counts on numbers of outbound messages to screens.
   4. Counts on communication lines traffic.
   5. Arithmetic totals for all like fields (dollar amounts).
   6. Placement of all totals in a compact table so they can be readily formatted for display and used for problem determination and restart.

B. Every data set should have a header which dynamically records the time and date of the last update, the program performing the last update, and the number of items currently in the file.

   Financial files should carry an extra date to indicate when the file was closed to additional input. The closing date and the date of last update will aid in balancing and reconciliation.

C. It is traditional in editing to build in special controls over big ticket, unusual occurrence, and catastrophe-if-wrong items.

D. If primary identifiers carry a check digit, if transactions are prenumbered at the source, or if prenumbered forms are used, the computer should check these descriptors upon input.

E. If programs are cataloged in several locations throughout a network, then each program should store a change level number as a constant in the executable code. This number can be used to verify the applicability of future changes and to guide the troubleshooting process.

In very large systems it may be impossible to apply a change simultaneously to all modules throughout the entire network. In these cases changes will be applied to nodes in some planned sequence and the system will have to be programmed to operate temporarily with programs in different remote nodes which may have different change levels.

137. Console Controls

Controls also need to be designed so that console operator actions are recorded and can be administratively reviewed. For example, it has been previously suggested in this handbook that one cell in the computer be assigned to count errors. As each operator message is constructed, the contents of this counter would be appended to the operator message. This establishes controls over the problem resolution process so the system administrator can establish a manual log in which he can record the symptoms, define the problem, describe any actions taken, and thereby create a problem history for later analysis.

Many systems provide senior console operators with special privileges which allow them to use utilities or other maintenance tools so they can arbitrarily change a record or a program to recover from some unanticipated failure. Although the use of such programs should be discouraged, circumstances will nevertheless occur when they are required. When the zap utilities are prepared, they must be programmed to leave an audit trail so the system administrators can review the actions taken to get the system going again.

138. Network Controls

A full-scale network encounters a complete set of control problems whenever it has some replicated data and some application program modules that are cataloged at more than one location, and a requirement for high availability (at least during certain critical hours of the day). To operate in such an environment, central management requires coverage programmers, central libraries, downline program load, remote program patch, upline debug, central records, and a full set of operator commands.

In addition, the central support staff must control each production program catalog; install a release number on every program and change it whenever the program is modified; and have the tools to correlate trouble symptoms by release number, environment, and configuration to identify duplicates and investigate patterns. Here are some other considerations important to central support staff:

A. The ability to display and print control tables of privileges, authorization lists, and command menus.

B. A library procedure flexible enough to handle:
   1. Routine operations
   2. Startup/shutdown
   3. Trouble diagnosis
   4. Program/system testing

   and a design which recognizes that logging requirements and privileges are different for each of the above situations, i.e., testing requires the ability to arbitrarily change status indicators so abnormal environments can be simulated.

C. The central support center should have definitions which establish classes of diagnostics and controls and procedures to determine when a situation requires diagnostics of a given class. The ability to diagnose and troubleshoot requires the capability to display and change. These may violate, or require the relaxation of, the normal operating controls established to maintain the integrity of the data base and its processing. Obviously these matters should be considered carefully, supplemented by full audit trails, and reviewed frequently to prevent abuse of privileges.

D. The central support staff needs the ability to declare a file or an entire data base as being under their exclusive control for diagnosis and maintenance. The use of this privilege must be logged and the central staff should be periodically reminded (by a timer) that a data base is locked so they do not inadvertently disable the system for an abnormal period of time.

E. The central support staff needs the ability to remotely change the selection criteria on data logging, to command that logged data is also to be transmitted to the support center, and the ability to rename members of program libraries. The staff also should be able to do a relink so diagnostic code can be temporarily inserted in the processing stream (or have the ability to temporarily patch production code so diagnostic snapshots can be taken).

F. Whenever the support center initiates long executions that inhibit access to system functions (data base load, unload, system startup, restart process, etc.), the local operator and the support center must be informed when these operations start and stop and must be able to interrogate the system to determine the status and progress being made. For example, if it takes 30 minutes to reload a data set, the data base containing that data set is probably locked for that period of time. Without the information described above, the local operator cannot answer questions from users and users cannot plan their work. Additional terminal inquiry transactions will result, thereby extending the blackout, as users keep trying to log-on to the system and access the unavailable data.

139. Training Features

When a life-cycle approach is taken to design, the attention given to several aspects of the system usually changes. One function frequently given increased attention is training. Many systems have been installed in which training was treated as a one-shot activity conducted at the time the system was implemented. Training courses are usually planned, definitive examples prepared, and coaches are provided during the initial system installation.

However, what happens when there is turnover in the workforce? What happens when a new site is added to the network? How are new features and enhancements introduced? What if there is movement in the user organization and whole activities are reassigned to personnel at a new physical location?

Usually ongoing training is left to the users themselves. Training materials are not maintained, examples may become obsolete, and since test data sets are seldom retained, the new operator goes live from the very first.

In small systems with low turnover this may be satisfactory. With large systems, training of new employees and retraining of existing employees as they return from leave or assignment or are transferred from unit to unit should be considered.

Previous items in this handbook have suggested that a few lines at the bottom of the screen be set aside, that some operator documentation be maintained on-line, and that users be provided with a HELP command so they can get information in context when it is needed. Other items have suggested that the users' skill level be maintained in the authorization table used at sign-on time and that this skill code be used to condition the system's response to the user. Heavy abbreviation should not be used in comments to unskilled users, while skilled users should still be able to use acronyms and codes to communicate with the machine and reduce keyboarding. Another item has suggested the implementation of a training mode so trainees are restricted from the global system commands which could cause trouble, and the provision of special training data sets so inexperienced users can use production programs on test data without disturbing the main data base. Some additional thoughts along this line are as follows:

A. The training mode should be designed to keep track of the training, the throughput, the errors, and any unusual command sequences so the trainee may be provided with an evaluation of his efforts at the end of the session.

B. Some systems provide selection from command menus as a way to progress through a command hierarchy, and others prompt users in an interactive dialog when data is desired from an inexperienced user. Almost all users appreciate these features when they are initially introduced to a system, but find such interaction laborious after some proficiency is gained. Thus some systems have provided a fill-in-the-blanks mode for experienced users.

C. Some systems have provided special commands to allow a second terminal to be slaved to the first terminal so a coach may follow the interactive dialog of his pupil without disturbing the training session.

D. For complex systems which support interactive searching, a library of proven search commands is frequently maintained. Thus the experienced searcher need not rekey a lengthy search command every time he wishes to invoke it.

These libraries provide useful training information that suggests how the system should be used for certain purposes. If provision is made to append comments to each of the cataloged search commands, a training coordinator can teach others the essence of the search and why it is constructed as it is.

E. The designers of one distributed system provided the ability to downline load training material whenever the system was changed and enhanced. Then as each operator reviewed the training material, their operator IDs were appended to a list. These lists were remotely accessible, and this allowed the training administrators at the central site to determine how large a population of users had viewed the training material and hence, when the new features should be scheduled for production operation.

140. Design for Performance

Previous items in this handbook have emphasized that performance is important in small systems since they saturate so abruptly, and since it is much more probable that a single application can dominate a small processor than in the case of large host systems. Performance has also been discussed from the human factors standpoint, particularly emphasizing the importance of maintaining keyboard rhythms for high volume data capture applications, and for maintaining a uniform response, independent of load, for all human endeavors. Here are some methods for achieving the desired performance:

A. A good designer will consider, and in some cases calculate:
1. The allocation of resources to the particular application.
2. Instruction path lengths and their frequencies of execution.
3. Estimates of application-to-application interference.
4. Capacity remaining on the system.
5. Execution times for individual transactions.
6. Performance while testing under full load with full data bases.
7. The degradation of performance with load.
8. Statistics and measurements to aid in the prediction of system saturation.

B. If the system will be moderately to heavily loaded, a designer should set response time targets for each keyboard function and execution time targets for each batch function shortly after the preliminary design is complete.

C. Assuming access to the data bank will determine ultimate system performance and throughput, a count of data seeks per transaction type should be produced shortly after preliminary design.

D. Determining requirements, hypothesizing systems, and sketching preliminary designs can be conducted almost independently of the specific hardware to be used for implementation. However, sometime during the detailed design process, the design team must become acutely aware of the I/O speeds associated with all the devices they contemplate using, and the effective transmission rates of all communications loops and circuits.

E. It is not unusual to assign an independent crew to build test cases from the detailed design specs. Such crews frequently go out of their way to build the test case libraries that correspond to the real world in scope and complexity.

If the data developed for functional testing is also to be used to predict performance, it will be necessary to choose the transactions so they are statistically significant and to establish the test data base so it has a known relationship to the production situation. Then the timings made during testing will be statistically significant and can be scaled up to predict the performance of the system under normal conditions.

141. Performance Trade-offs

In performance work some trade-offs are possible:

A. Excess CPU cycles can be used for packing and unpacking data records to disk and messages to be communicated.

B. Extra memory can be used to block disk records or communications messages and therefore reduce the number of I/O calls and the communications line time. (Warning: Excess blocking complicates restart.)

C. If file space is available and processor cycles are in short supply, records can be held on disks already formatted for display. This gives quick response to terminal input since the CPU can merely fill in the blanks and transmit the response with no need to construct the entire response from a packed format.

D. In some cases the user's tolerance for delay suggests that an entire multi-screen response should be completely formatted before the first screen is offered; then the user can page forward and backward at will without being inhibited by the processor's need to format subsequent screens.

In other cases the user is vitally interested in the first screen of information and can be expected to contemplate it for some time before requesting additional information or inputting new commands. In this case the process sequence should be biased to present the first display screen as soon as possible and either queue the processing of the remaining screens or delay the processing of subsequent screens until they are requested. (The choice depends on the human factors of the situation.)

142. Performance with Growth

Sometimes a processor is ordered with a minimum configuration. The design team then knows that careful design and tuning will be required, otherwise the hardware on order will be unable to carry the necessary load. In these cases the designers will make extensive use of reentrant programming so programs and edit tables can be read from disk without being restored.

Data records may even be laid out so the data fields are ordered in increasing degree of volatility. Then only the rightmost fields must be recorded as they are (usually) the only ones that change during processing.

The placement of data sets on disk is usually of no concern to the designer. However, if every unit of capacity is being extracted from a hardware configuration, the sequence of disk seeks can become important. In some cases, reducing the seek time slightly doubles the throughput of the disk since latency revolutions can be saved in critical seek, read, process, write sequences.

If careful disk allocation is required, the track assignments for data, application programs, and system purposes should all be reviewed.

143. Performance Measures

If multiple applications on a single small computer are contemplated, and if it appears that the computer will be heavily loaded, each application design team should receive a ration of machine resources and attempt to prepare their code so it runs satisfactorily within those resource bounds. This will require the programmers to define the target workload carefully, to quantify the system resources used as the design is set down, and to measure the resources actually used after the code is written. With experience (and good recordkeeping), reasonably accurate coefficients for performance estimating can be determined.

Some performance techniques from large systems also apply to small processors. An initial set of measurements might consist of:

A. External message activity.

B. Buffer usage.
C. Internal message activity.
D. Transaction processing rates.
E. Actual idle time (cycles) remaining under various loads.
F. Specific message traffic and I/O counts.
G. Max/min buffer usage.
H. Processor utilization.
I. Max/min queue lengths.

144. Design for Testing

Statistics have shown that about half the cost of development in a large computer center is spent by the time unit test is completed and integration testing is about to start. If a small system is dedicated to a single application, and if that application does not tax the installed capacity, one should expect some savings in development test expense. On the other hand, if the programs are being prepared for use in a variety of locations with a wide variety of environments and workloads, test expense might even increase. Here are some suggestions which tend to reduce the cost of testing:

A. Applications should be designed with testing in mind. Specifically they should be modular and not monolithic. Further, each application should respond to the setting of a test mode indicator so it can be remotely placed in a test mode, driven by one or more transactions, and have the results routed (as a result of being in the test mode) back to the person submitting the test. Further, since many production applications will keep counts of errors uncovered during edit and processing, the test mode bit should inhibit the debug run from contributing to these error counts and related alarms. This way, the operators will not be alerted when a few test cases are run through a production system.

B. If remote testing is to be supported, the programs which manage queues at the main interfaces must be accessible so

the normal queueing and routing schemes can be overridden for test purposes.
C. For full-fledged remote testing, a test control module will be required which will stand astride the interface between two modules when it is invoked. Thus the control module can record the inputs and outputs to each major processing module as the process sequence occurs.
D. If applications are developed centrally, and if sufficient storage space is available, the inputs, working storage, and outputs from each module can be left in accessible space so a snapshot between modules will provide a before-and-after picture.
E. When testing a big system, test cases must be identified and test quality must be tracked. The following controls will be required:
 1. Know what test cases have been run against which modules (and with what result).
 2. Know when programs have been compiled or recompiled.
 3. Know the thoroughness (not quality) of a program's tests.
 4. Know the extent that tests have exploited the range of variables, the combinations of mid-range variables, and the combinations of extreme values that have been tested.

145. Incremental Conversion

Few, if any, really new applications seem to exist in commercial data processing. In most cases systems development is involved in improving an existing application to add functions, reduce cost, or improve performance. Therefore, the systems designer should consider seriously, early in the design process, the migration problems he may encounter.

Most (user) managers prefer to install systems incrementally so experience can be gained in using the new system before the entire operation is committed to it. In addition, people must be trained, files must be converted, and if an on-line system is being introduced for the first time, external procedures must be adjusted or the benefits of the on-line system will not be achieved.

Thus a systems designer might decide early in the design cycle to build a system which will run in parallel with the existing system. Then after confidence is gained with the new system, the old system can be discontinued. To support this strategy it may be necessary to add features to the new system so it can emulate the old system and the existing flow of paperwork.

In other cases a designer may find specific functions of the existing system which can be automated in their entirety and cutover as the first increment of processing. If a systems analyst finds that an existing clerical operation uses hard copy printouts or microfiche generated by a host system as reference materials for answering inquiries about product status, consider the following sequence:

- A reasonable first step would be to install a small computer with a local printer. The hard copy listings then could be routed from the host to the local computer for printing (thus avoiding courier runs or transportation delays).

- A reasonable second step would be to hold the information on the listings in a reference file and provide on-line inquiry into that file.

- The third step would allow the capture of change data from local terminals.

- The fourth step would be local update and transmission of batches containing those same update transactions to the host system for overnight processing.

One large organization has all of their recordkeeping on a big, monolithic 10-year-old batch computer system. This system is so vital to operations that abrupt changes are unthinkable. To assure that conversion and migration are given adequate consideration, the development project is organized with a migration manager who is coequal with the systems architect and the programming chief.

While the previous sequence of activities may not be appropriate to all environments, it nevertheless demonstrates a way to incrementally convert from an existing host system onto a distributed system while maintaining the confidence of local management, and providing increments of change that are small enough to be assimilated by the clerical workforce.

146. High Availability

Frequently this handbook has addressed aspects of restart, recovery, and availability. Modern hardware is more reliable than previous families of equipment, but on-line interactive network environments are more demanding. Techniques have been presented that reduce the terminal blackout time following a failure by providing features which make problem determination easier, which facilitate restart, and which provide for partial file lock and selective reconstruction so on-line processes may be restored even though the full data base may not be available.

Previous discussions have described organizational responsibilities at both remote and central sites and have sketched some of the documentation and procedures that must be available at central sites if they are to provide remote support following system failures.

If an applications systems designer cannot live with the inherent reliability/availability in his computer system, then he must be careful to configure the system and design the programs to achieve a higher availability than that natively available. At the risk of stating the obvious, one observes: High availability does not automatically happen!

The native availability of the computer system is what the manufacturer sells and the user buys. This is the minimum systems availability for the manufacturer's specific design, architecture, and circuit technology. This is also the minimum cost system obtainable from that manufacturer. If additional availability is required, additional equipment and programming will be required, over and above the minimum.

High availability must be designed in by a creative mix of hardware, software, communications, applications, procedures, people, and facilities. Sometimes this can be done without changing a vendor's basic offering. However, it may require extending the vendor's offering in ways he has not planned for.

A partial list of problems which must be addressed when designing for high availability are:

A. Reconfiguration after an outage.

B. Rereconfiguration after a repair.

C. Restart and recovery within the maximum blackout time available.

D. Diagnosis and repair, after the system has been reconfigured to carry the mandatory production load, on the surviving hardware.

In batch systems, if the failing device can be pinpointed, the shift supervisor can decide whether to reconfigure around it or call the system down until it can be repaired. He is influenced in these decisions by the availability of competent vendor field personnel and spare parts, and the pressures from his current workload.

In on-line systems the shift supervisor's inclinations are different. First he needs to know which device/component is failing. Second he needs to know whether he can continue to provide service in a degraded mode if he reconfigures around the failed device. Third he proceeds to reconfigure and restart the system so he can continue to offer service to his on-line users. Then he worries about repair.

Thus the batch manager is concerned with the efficiency and efficacy of his computer shop, and the on-line manager worries about continuing service even though it may be in a degraded mode.

Historically the CPU was the system component most prone to failure, so a mystique has built up over the years about cross-connecting dual CPUs to a shared disk to enhance availability. Of recent years improved electronic reliability has greatly outdistanced the improvements in electromechanical reliability. Thus a person attempting to design a high availability system must perform a failure analysis on all the critical components of his system (hardware, software, applications, and personnel), to determine which components are most likely to fail. An estimate then should be obtained of the time required to diagnose and repair the typical worst case problem. Given the results of a failure analysis, a designer can then proceed to configure hardware, adapt software, design applications, and train personnel so the likely worst blackout is still within the allowable constraints.

A partial list of items to consider when designing restart follows:

A. Authenticating who is at the terminal and re-establishing session security.
B. Continuing a session based on existing context or reinitializing the session to some past set point.
C. Determining the status of all active files, and then backing out transactions, restoring previous versions, or partially locking the file set as appropriate.
D. Performance is a perennial problem. Restart activities plus continuing production plus the problem determination-diagnosis-repair-verify sequence all must be performed on a degraded configuration. This is usually the worst case load.
E. Many designers omit features from their designs which will keep the users and the system managers informed as to the magnitude of the problem, the elapsed time the system will be operating in a degraded mode, and the time the full system will be back into production. The data for these estimates, as a minimum, must be produced by the degraded system so it can then be broadcast to all users. The users then can plan the remainder of their work day accordingly. At the same time the system managers must be informed of the system status so they are prepared for contacts from irate users.
F. Graceful degradation trees and recovery sequences must be designed in. While the two routes must have common end-point nodes, the recovery route may not be the same as the degraded route.

For example, if a disk drive failed, the active file set would be rebuilt on another drive. There is no reason to restore that file set on the original drive (and incur a double move penalty) when the original drive is repaired unless contention and performance degradation is severe with the active files placed on the alternate drive. Even if performance degradation is severe and the file set must be moved back, the system manager should be able to choose an auspicious time for the second move rather than impacting the on-line users twice within a very short elapsed time period.

147. Degraded Operation

A prudent system designer would design for continuing processing at his node even though some components of the local system or the network may be unavailable. Some techniques for building survivable applications involve the following:

A. A unique system status indicator should be stored in the node, having three states: the full system is in operation, the communication line is serviceable but the host computer can provide only a defined subset of its normal functions, or the communication lines and/or the host computer are completely unavailable.

B. The operator at the node should set this system status indicator properly and reset it whenever conditions change.

C. When a transaction is received from a terminal, the application program should check the system status indicator to determine whether the transaction can be processed routinely or if one of the abnormal processing modules should be called. If the host system is down and if host dialog is absolutely required, the node must either reject or queue the transaction.

   If the transaction can be processed satisfactorily at the node, and if that processing normally sends a message from the node to the host so two data bases can be kept in step, that message can be queued in the event the host is down or available for inquiry only.

   If the normal processing flow requires the host to validate a transaction or authorize an activity, sometimes the application may allow this validation to be selectively omitted without undue risk. In other cases the host communications can be ignored altogether, provided the proper conditions exist at the node, i.e., a space available assumption can be made if an air reservation is being booked more than two weeks in advance (or a transaction causing deferred confirmation can be generated).

D. The two extreme cases are easy to handle. With the entire system in operation, the process runs normally. With the host completely unavailable, any transaction requiring mandatory dialog is rejected. The intermediate cases remain to challenge the designer.

   If communications with the host consist of secondary transactions to update a data base, then these can be queued if the host is down. However, when the host comes back up, the queue of transactions must be processed as a minibatch. If an error is found in one transaction, while the minibatch is being processed, the normal error dialog cannot be conducted if the terminal operator has completed his shift and gone home. Thus the processing of a queue of deferred transactions must provide for alternate routing of error responses to a supervisor's console.

   If routine processing causes the host to return an information message to the node, then the alternate routing program must send these acknowledgement/approval messages to the supervisor's console so the routine closeout messages can be ignored and any abnormal advisories can be acted upon.

E. In some applications a full function computer system will be mandatory to support a work unit. Such an office is severely discomfited if the local node starts rejecting transactions for any reason during the normal business day. However, unless the system has been designed for high availability, some critical portion of the node, the communications, or the host systems may from time to time cause routine processing to be interrupted.

   Under these conditions the prudent application designer will have provided some necessary manual activities so the workforce can be productively employed while the system is being repaired. (In the design of the global system a small backlog of noncomputer-related tasks can frequently be maintained in each office. There are manual records to keep, forms to file, training manuals to be mastered, and the like. Thus if the computer system goes down infrequently, the office staff can be directed to a backlog of training and education materials until the computer comes back up.)

In other cases pads of manual input transaction sheets can be retrieved from prudently stocked supply cabinets and the input section of each form (looking suspiciously like a keypunch transmittal) can be completed. Ideally these forms would be designed with the same graphic layout as the CRT screens used by the operators for routine input. Thus instead of keying data into the screen, the data is written in the appropriate fields on the form.

Then when the computer comes back up, the forms can be keyboarded as a batch and if any of them are rejected, the rejection messages can be copied from the CRT and entered on the bottom of the input forms, thereby generating a "cause and effect" document which can be manually resolved and later reentered into the computer system.

F. The system designer preparing for all contingencies will note that office operations following a major failure have a different flavor from routine operations in that same office. If they are different enough, the designer should plan for such contingencies and build a catchup mode right into his application system. Thus in routine service the computer would see single transactions in more or less random streams from the various on-line terminals, whereas in the catchup mode the computer might see fewer terminals logged on and batches of similar transactions being entered from each terminal.

In the former case the best service could be given by dispatching transaction processing programs one at a time as the transactions are received. In the latter case one might capitalize on the fact that the transactions have already been batched and use batch processing techniques to minimize the total wall clock time for processing the entire set.

Consider the case of the computer that fails late in the day, causing the office manager to hold the most proficient operators on overtime to enter the backlog of transactions. In this case the normal dialog should be curtailed and the batches should be processed as efficiently as possible. As a result, the employees will get home sooner and the overtime will be minimized.

148. Restart, Recovery, and Availability

In addition to the hints given earlier, here are a few more related to restart, recovery, and availability:

A. If reliability/availability controls are required:
1. Selectively inhibit message creation in the face of overload.
2. Remove message inhibitions by class and source.
3. Temporarily inhibit polling.
4. Temporarily reroute messages.
5. Temporarily queue output for delayed display/print.

B. For fast restart, a second copy of each local system's configuration and status table must be dynamically maintained at the network support center.

C. In environments supporting both on-line and batch, keep the batch transaction processing capability intact from the development project. Document this capability and use it for emergency backup when the network is down.

D. On a two disk system, put the transaction log on a different disk from the master file to level the I/O load and to increase the probability that the loss of master file and restart log will not occur simultaneously

PROGRAMMING

149. Design Decomposition

The main-line design process is not completed when the detailed design specification is published. Following the review and approval of that spec, the designers and the programmers must work together to decompose the design into modules which can be individually written, tested, and integrated. Decomposition is an art, producing large savings in development time and dollars if done correctly. Unfortunately it is an arcane art, ignored in the popular literature and uninvestigated in academic circles.

Van Slyke was quoted earlier, when he stated that systems were decomposed because the parts were more homogeneous than the whole. Systems are also decomposed to obtain units of work that are within the capabilities of the available programmers, to establish the boundaries of modules so growth and change can be easily accommodated, to establish intermediate level boundaries so the system can be efficiently repackaged for multiple environments and workloads, and to structure the application programs internally so they are easier to test.

A few decomposition hints follow:

A. No general theory is available to guide the applications designer in splitting up excessive processing sequences. However, one specific instance occurs frequently enough to warrant discussion. Most transactions progress through the following processing stages:

1. Field edit
2. Record edit
3. File edit
4. Processing
5. File update
6. Response generation
7. Response formatting
8. Response presentation

Processes 1 and 8 can be performed in the node nearest the terminal. Processes 3, 4, 5, and 6 can be performed only at the node with access to the data base. However, this leaves processes 2 and 7 which can be performed optionally either at the originating node or at the data base node. These two processes allow for a small portion (maybe as much as 10%) of the processing to be optionally assigned to one node or the other to achieve load balancing.

In the case of a data capture application, only processes 1, 2, 3, and 5 are required to capture the data, edit it at three levels, and queue it for deferred processing. When designing a data capture application, recollect that process 2 can be executed wherever available processing power exists. All that is required for its execution is the transaction itself, some edit rules, and the cataloged application program.

B. Another approach to decomposition involves identifying the common and unique functions, determining how many unique functions can be made common with the addition of a customizing parameter and table driven code, and then analyzing how the large functions can be intelligently broken up.

Every separate module should have a formal interface to define the call and the parameters to be passed at the time of the call. Similarly all error conditions should be defined along with all meaningful data that is returned with the error.

If any function requires I/O, the performance-sensitive designer will determine if the request can be isolated from the processing so the data can be requested early, allowing I/O to be overlapped with as much processing as possible.

Finally after the entire application has been decomposed, the designer should trace the path taken through the modules for each transaction to assure that performance and throughput are likely to be acceptable.

C. For years software designers have been talking about device-independent programs. They have successfully provided systems so the application program is not wedded to a specific device and its unique characteristics. But the device *type* still shows through. Thus it will never be possible to make a serial tape look like a cyclic disk. While a cyclic disk can be made to logically appear like a random access memory, the performance of the device is still dominated by the cyclic storage media, and the response time to a random stream of addresses is affected by the rotating cycle.

Thus when decomposing code, the major hardware subsystem boundaries should be respected. A clean interface to the data management system, a clean interface to the terminal subsystem, a clean interface to the communications network, and specific narrow interfaces to one-way devices like card readers and printers should all be respected. While decomposition can be separately explored within each module and within the various parts of the application code, it would be a mistake to package functions from both sides of a major interface in a single module.

TEST DEVELOPMENT

150. Test Tools

Previous items in this handbook have discussed structuring working storage so snapshots are more meaningful, putting controls in applications so testing can be performed remotely, and building measurements and statistics into application programs which will initially support testing and later support production operation. Items have even described table-driven code and specially formatted displays which will make programs easier to test since they are more disciplined and easier to construct and understand.

Some items follow that should be considered for inclusion in test tooling:

A. Table-driven driver programs which generate test cases from parametric inputs.

B. Dictionary driven dump formatters.

C. Standard statistical tools to help find patterns of occurrences in long runs.

D. A regression test data base containing:
   1. A library of test cases.
   2. A library of test procedures.
   3. A library of test output.

E. For voluminous tests, a regression test system which directs output to tape/disk and compares it against previous output to detect changes.

F. A test control program built to drive modules which in turn simulate keyboards entering transactions. Use the test control program to provide input from one terminal in a consistent manner or a whole set of terminals as a load test.

G. A logging program built to record all traffic to and from a terminal for later analysis (time stamping all records), and an analysis program to print the records or use the messages for dialog analysis.

A special processor program can use this recorded exchange to build a test data base, or as input to the test control program for load or regression tests.

H. Given a data base and a terminal dialog, use the data base activity log to subset the data base so it contains only the active records that correspond to the transactions.

151. Blanking Buffers

To enhance snapshots, compile in a test module which zeros out the input data buffers prior to accepting a new transaction. Then you can be sure the snapshot contains only data that was received with each new transaction, and you need not be concerned about the residue from previous transactions.

152. Debugging Production

Good practice dictates that check digits on identifiers be verified when transactions are received and verified again before items are added to the data base. Other checking modules are frequently available during test, but are discarded when the system goes into production to save space, increase performance, or both. If space is available, consider making these checking modules optional, so a single parameter can cause them to be reinvoked at any time. While they will waste a few cycles during routine production, they will more than pay for themselves if the system becomes unstable and troubleshooting is difficult.

153. Design for Testing

Efficient testing is frequently the result of application design choices made to accommodate testing and of test tools which exploit those choices. Consider the following:

A. If all application dialog with user terminals goes through a single module and only that one module calls the terminal handler, then load tests are easily accommodated by replacing the terminal handler with a test driver-recorder module.

The interface would be the same as before and all applications code under test would remain unchanged.

B. If the application is internally structured with formal interfaces, and if the control flow across those interfaces goes via a monitor module (so the process is interruptible and so debugging and troubleshooting are accommodated), the monitor module can be modified for test purposes to capture and/or modify the output from any process before it is queued for the next process.

C. If an application selectively writes input transactions on a log, then a parameter change can cause 100% of the input transactions to be written on the log. The records from the log can then be processed by an extract program, so data for volume test can be constructed by replicating the sample of live data obtained from the first few operational terminals.

D. One company had extensive experience in building, modifying, and tuning on-line application programs for a network of small computers. To perform systems tests without impacting production, they built a test system which runs on their host computer and employs utility programs to set up the test environment, a terminal simulator which can be driven by a data set, and a series of logging and debugging tools to track the outcome of the tests. In addition, they constructed performance measurement tools which record test results and then make allowances for the test environment. Thus they could predict system performance when the programs under test were running live.

154. Flawed Tooling

Sometimes test tools require special analysis and design to be useful. Two flaws appear in test tools with sufficient frequency to warrant mention:

A. A pair of tools is frequently designed for regression testing. These are used after every program change to prove that the modified system produces the same outputs in response to a stream of input transactions as the previous version of the system.

Although the stream of input transactions may be the same, differences in priorities, queues, and timing can cause the sequence of results to vary even though the sequence of input transactions was identical. Thus the second of the two programs will often incorrectly indicate unsuccessful results if attempts are made to compare responses blindly, one-for-one between the new version and the old.

B. In some very large systems a terminal simulator is constructed which replaces the terminal handler module and drives the transaction interface with a primary input stream of dialog.

To get a true test, the operator's delays for keying and assimilating outputs (think time) must be simulated. Further, the randomness in arrival times among several input transactions must be maintained. To do less is to preload the system queues with an unusual burst of activity and to keep them saturated in an unrealistic fashion.

155. Performance Tooling

If a single small computer is to support multiple applications, and if resources are preallocated to each application to minimize interapplication interference, some test tools are in order to measure the quantity of resources remaining while the first application is running a load test in a simulated environment. One procedure would be to devise a general purpose table driven program with parameters set to arbitrarily use all resources not assigned to the program under test. This would deny any extra resources to the program being tested so any performance degradations would show up in the tests and not when the second or third applications were being installed some months later.

An alternate version of this program would allow the program under test first priority access to the resources it needed and have the test environment simulator exploit the rest, measuring how many disk seeks it was allowed, how many messages it could transmit, and how much storage it could retain. In this version the simulator would report capacity remaining.

Either version would be equally useful for the first application. The second version would be more useful after multiple applications had been installed and were running simultaneously, as it would measure the residual capacity available for growth.

DOCUMENTATION

156. Documentation Strategy

Documentation has been mentioned in several previous items. To recap briefly: Prepare a preliminary user's manual as soon as preliminary design is completed. Prepare a high-level systems overview as part of the systems analysis effort and modify it as required following preliminary design and detailed design to keep it current and available to management, auditors, or employees newly assigned to the project. As soon as detail design is completed, prepare user's scenarios which describe the man-machine dialog in detail and indicate the expected elapsed time for each automated and manual processing step.

Eventually training materials will be required for every person associated with the system, and manuals which address both automated and manual tasks will be needed for all skill types. For instance, training and manuals would be required for the central support center, for console operators at remote sites, for users, and for user management. The user's manuals would probably contain several sections on the system itself with optional chapters inserted for the applications installed at each site.

Within the chapters addressed to each application, there would be sections providing an overview of the application plus detailed sections that covered both routine and abnormal operations. There should be sections discussing the manual processing that takes place before and after the automated portions of the application.

In small systems this documentation can be prepared with a typewriter. With larger systems the complexity and variety of the documentation will require at least a word processing system and perhaps a computer-based text processing system which will support global edits and typesetting. If some of the documentation is placed on-line to be available in response to a HELP command, references in the on-line documentation will be required so the abbreviated HELP commands can cite chapter and verse in the hard copy documentation.

User's manuals are always a chore to write and good ones are usually produced by persons who have special technical writing skills. If a designer is contemplating the implementation of a network with geographically dispersed nodes, he should not underestimate the magnitude of the documentation chore.

As part of the detailed design process, the design team should produce the list of the manuals to be prepared, a table of contents for each manual, and an estimate of the number of final pages required to cover each topic for the intended audience. If previous documentation is used as a guide, the designer of the documentation series should be sure to adjust the actual page counts from existing documents for any differences in profile between the existing and intended audiences. Manuals prepared by DP specialists for professional keyboarders can be quite concise and explicit. If, instead of well trained keypunch operators who have ten years of experience, the manual is intended for receiving clerks who have never before encountered a computer terminal, the format, style, content, and rate of delivery will make the manual larger and in some cases force a change in style.

After an inventory of the manuals to be produced is available and the page counts are estimated, project management can assess the magnitude of the documentation effort and review the staffing and selection of tools to assure their continued applicability.

TRAINING

157. Training Strategy

This handbook has previously noted that a large system requires the training of a large population, and that any distributed system requires the training of several kinds of people to perform different roles. Many of these training requirements are familiar to the experienced designer, although the distributed system environment places a different emphasis on some aspects of this training. Training users to perform their own input and inquiry directly may be a familiar process to the designer who has been involved in user-oriented on-line systems. However, training user management to assume the new responsibility for systems operations which are delegated to them is unique to distributed systems and will require some original thought.

Similarly, training essential support staffs will require some thought since the breadth of roles assigned to personnel within the support center is unusually large, covering hardware, software, applications, and system administration.

One should not overlook the training needs of the design team itself. Even if the team is made up of seasoned analysts and programmers, they may lack experience with small computers, with communications networks, and with user personnel who have little or no schooling in data processing.

The training needs of several levels of user management should also be recognized as potentially new and unique. The customers of a large central data processing complex are frequently familiar faces. If the distributed system is to truly serve the needs of the remote user, and if those remote users are to participate in the design of the system and the manual procedures that surround it, they may need a short course in data processing.

If their training needs are ignored, they will still learn rapidly during the Definition and Design phases of the project and will apply this newfound knowledge towards changing the requirements. To get *informed consent* from these users, they must understand their present environment, some DP principles, and the proposed system before they are asked to approve design decisions. To do less is to run the risk that they may overlook something. For example, they may forget to mention a query type so the design team would later find out it is missing a secondary index or even some data elements in the master data set.

The analysis matrix technique can be used to define training needs. If the various personnel who require training are listed down the left side of the matrix, and the various subjects to be covered are listed across the top of the matrix, notations at the intersection of a row and a column can indicate the level of training a given individual requires in each specific subject. Rather than placing Xs at the intersections to indicate some undefined need, the designer should code the level of knowledge required and the type of media (hard copy manual, self-study with visual aids, formal training course, automated interactive instruction, or coach and pupil dialog) that can best convey the subject to the student. After detailed analysis, a third annotation can be made to indicate the number of instructional hours required to cover that topic appropriately for the specified audience.

The designer should be careful not to concentrate exclusively on initial one-shot training needs, but to assess ongoing needs due to turnover, transfers, leaves of absence, and major reorganizations.

CONVERSION

158. Conversion Strategy

People, files, and physical facilities must be converted before a new system can operate successfully. If people are scarce and the workload is heavy, spare time may not be available to devote to training. Thus temporary staff or overtime may be required to give the present staff enough relief so they can start to assimilate the new system.

If the current system is automated and if the goal is merely to migrate existing system functions outboard, file conversion chores may be minimized. However, if the data resides in manual files or if automated files lack key data elements or if the integrity of the existing automated system is not satisfactory, file conversion tasks can vary from significant to formidable.

If the user occupies a crowded space or if there is equipment already installed that occupies a crowded space, there may not be enough room to install new equipment. Further, the installation of new equipment and its cabling usually requires some disruption to the current operation. Depending on the company and the current situation, getting a few hundred square feet near stable power in a benign environment, with good access to overhead or underfloor cable runs, may be a time consuming political chore.

If a system exists and if that system is vital to the daily operation of the work unit, the management may not allow an entirely new replacement system to be brought in and installed in a single series of events. Despite the experience of others, they may not trust the equipment, the software, the development team, or the schedules. If this is the case, then an installation strategy requiring complete cutover in a short period of time may be unacceptable. The basic design may need rework so a system can be installed in increments and still provide useful function between the installation intervals. This will allow managers of previously unautomated functions to phase the introduction of the system into their operations and stretch out the time for user training and systems assimilation.

Although you may be installing a system which has sophisticated communications protocols, can emulate the terminal system you are replacing, and has hardware and software designed for easy installation, the main part of the conversion chore remains to be faced by the design team. How much change can the user tolerate, how fast can he absorb increments of change, how does the system operate between those intervals, how many versions of the documentation and how many training short courses will be required? Can all of this be accomplished so any one set of changes can be backed out and the previous stable operation restored in the event serious trouble is encountered?

In addition, very big systems will tax the ability of the development project to install them. Even with the most ornate preparations, several hundred processors will be a chore to install. Even if the programming were perfectly done, obtaining communication circuits, resolving problems with physical facilities, and training a thousand or more people is a formidable undertaking. Therefore practical considerations dictate that multiple simultaneous installations be deferred until a few pilot installations have been established, load tests have been performed, training materials have been revised, and an articulated installation plan has been prepared. This may result in a carefully phased cutover extending from one to two years with oversize HELP groups being required to support the varying levels of user skill and systems maturity until all systems are installed and settle down.

PHYSICAL FACILITIES

159. Site Planning

Physical facilities for a small system are not usually a major concern, but cannot be ignored completely. From the installation manuals available and the preliminary description of the hardware configuration, a site questionnaire can be prepared and sent to every user manager who will eventually carry the responsibilities of remote site administrator. When the responses are processed from such a multi-site survey, most of the administrators will have found adequate space and a few will have problems.

For those who were successful in finding adequate space, an installation planning checklist can be prepared so they can claim the space and start any basic renovation that may be required. For those administrators having trouble finding adequate space, a series of phone calls will resolve some of the problems, and management action or a visit from a facilities planner will usually resolve the rest.

While most small computers do not require much more than a normal office environment, they do require clean power, cable access for local loops, telephone access for communication circuits, physical access during all shifts it is expected to operate, and security access controls appropriate to the level of information being processed. Any electronic system also benefits from constant temperature within the comfort range, a dust-free environment, and a nominal amount of soundproofing. A small-sized machine room would require about 270 square feet to provide space for the equipment, supplies, and some working space for transient programmers. Additional space would be required for a terminal area, resident programmers (if any), and for medium to large configurations.

If the system processes precious information and requires encrypted communication lines, this probably dictates full access controls and a locked room.

INSTALLATION

160. Physical Installation

With a little planning, the physical installation of a small system can be readily handled by local personnel. After the room is chosen, a floor plan prepared by the local administrator will allow a member of the design team to draw an initial physical layout. If the planning is centralized, only the first few of these layouts will need to be verified by vendor support people prior to sending them to the field. An installation checklist can be prepared from vendor manuals and tested on an initial installation. Checklists can then be sent to all site administrators so the equipment can be received, unpacked, set in place, and cabled. Some small systems even come with running software so semiskilled DP people can bring the system up and perform initial acceptance tests. Following these initial tests, these same people will be able to follow a tested script provided by the design team to perform basic tests on the entire configuration.

If all the tasks described above have been successfully accomplished, the system should be ready for operation when the project's installation team arrives to install the program sets, bring up the files, and exercise the communications circuits.

5
PROGRAMMING PHASE

Figure 5.1 combines the task phasing and the activity list for the Programming Phase. The principal active tasks during this phase are Programming and Conversion.

All the design hints given in this chapter may not apply in every case. But even a hint that does not apply may suggest a previously unrecognized area for study or investigation. It is suggested that each reader have a notepad handy as items unique to his environment may occur to him while reading the hints provided. If all personnel at an installation pool their notes, they will have augmented this handbook with items specific to their environment and class of work. Thus designers that follow will benefit from those who have gone before.

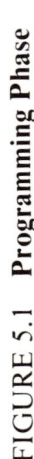

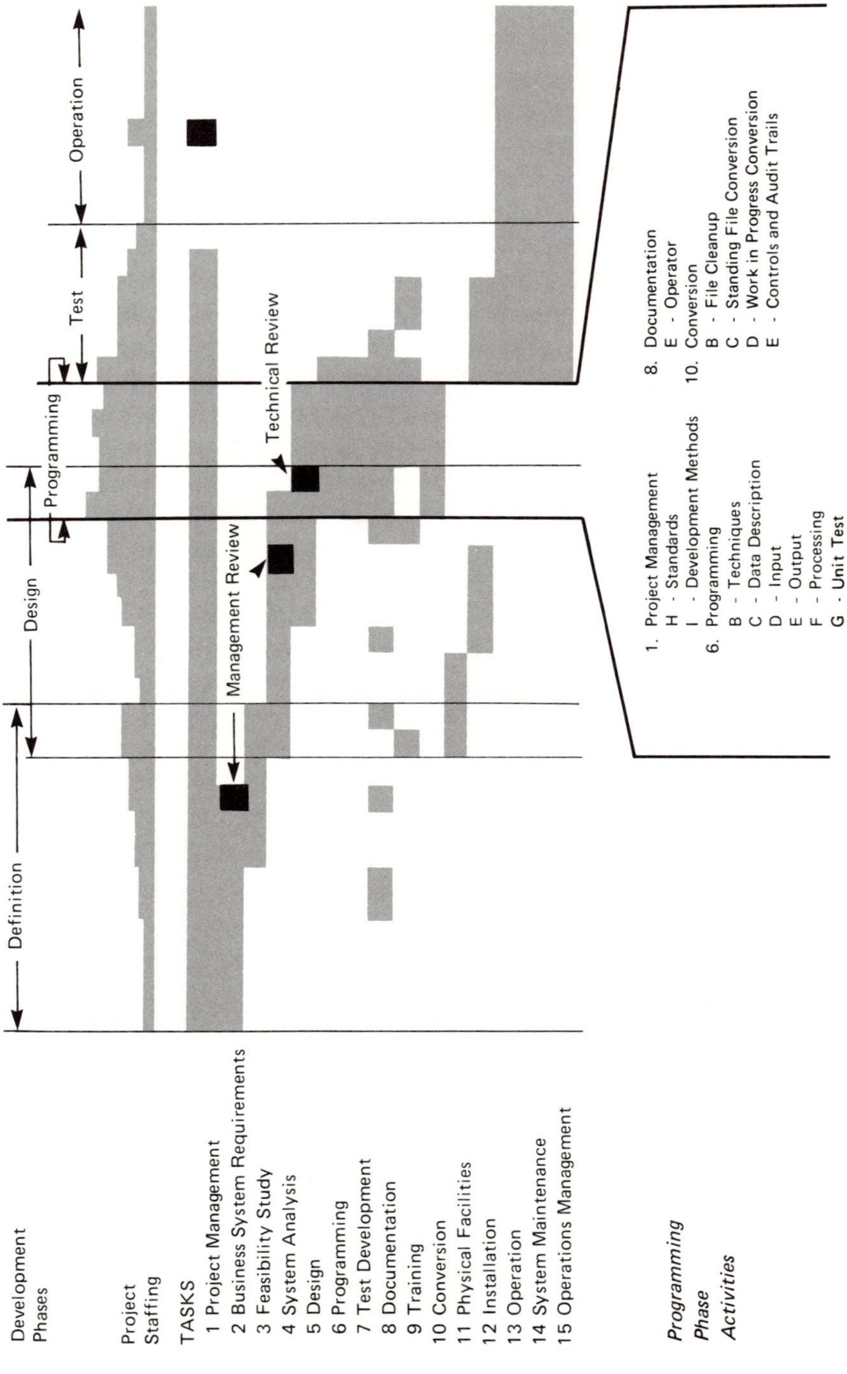
Development Phases
Definition
Design
Programming
Test
Operation
Project Staffing
TASKS
1 Project Management
2 Business System Requirements
3 Feasibility Study
4 System Analysis
5 Design
6 Programming
7 Test Development
8 Documentation
9 Training
10 Conversion
11 Physical Facilities
12 Installation
13 Operation
14 System Maintenance
15 Operations Management
Management Review
Technical Review
Programming Phase Activities
1. Project Management
H - Standards
I - Development Methods
6. Programming
B - Techniques
C - Data Description
D - Input
E - Output
F - Processing
G - Unit Test
8. Documentation
E - Operator
10. Conversion
B - File Cleanup
C - Standing File Conversion
D - Work in Progress Conversion
E - Controls and Audit Trails

FIGURE 5.1 **Programming Phase**

PROJECT MANAGEMENT

161. Replanning the Project

If the project follows the schedule presented on Figure 5.1, the beginning of the programing phase follows the second Management Review. Thus the requirements have been determined, the preliminary design has been published, a detailed design has been prepared, and the magnitude of the conversion problems has been assessed.

If the file conversion or the training problems were severe, they must be given special attention so the tasks necessary to accomplish file conversion are itemized, their dependencies are noted, and the amount of effort to accomplish each task is estimated.

As programming is about to start it is probably appropriate to review the schedule and the staffing. Is the sequence of events still reasonable, or has the Design phase uncovered more work than was originally anticipated? Is the staffing still reasonable now that the analysts and programmers have been trained on the hardware and software they will be using?

After a brief review, if the previous plan looks unworkable, the project manager should replan the effort, using the same techniques described in Chapter 3 and, based on the availability of headcount and talent, reschedule the remaining effort.

The schedule for the Programming, Test, and Operation Phases is vitally dependent upon the quality of the planning, the thoroughness of the design effort, and the competency of the project management. Frequently programmers start to work only to find that the design lacks quality. While it is not unusual for some detail design to be completed in parallel with programming (as shown on Figure 5.1), design changes or major additions to the design usually foretell schedule slips.

Careful measurement of progress during the initial programming period will allow the quality of the design to be ascertained and will alert management to possible schedule slips.

If the design is complete and if it has been properly decomposed into modules, then the programming staff can proceed to describe each module in detail, program it, and test it. However, if during this programming process many new modules are uncovered, schedule trouble is sure to follow. A technique known as Rate Charting* will allow a project manager to track progress across this very critical period.

162. Review Programming Standards

Previous items in this handbook have urged the establishment of programming standards so the results of each development project have a degree of uniformity which will ease maintenance and training. Standards will also ensure that programs are internally constructed so they maintain the appropriate degree of flexibility, modifiability, and portability to support their intended environments. As the Programming phase starts, the programming standards adopted should be reviewed for completeness and applicability. If any other development projects have been recently completed, those experiences should be distilled to determine if the body of programming standards requires modification. Finally, the project manager should verify that the programming crew is aware of and properly schooled in the standards they are to use.

PROGRAMMING

163. Programming Kickoff

Assuming the design is complete, the programming team is trained, and one or two programmers have actually coded and checked out sample code, the programming team is still not ready to start unless they can answer yes to all the following questions:

A. Have the designers prepared a briefing which discussed the human factors concepts and the features of the design explicitly aimed at achieving satisfactory man-machine dialogs?

---

*Terry R. Synder, "Rate Charting," Datamation Nov. 1976, p44.

B. Have the designers prepared a briefing and discussed the features of the design explicitly included to provide for growth, portability, and remote operation?
C. If multiple applications will be installed on a single processor, does the programming team understand the resources allocated to the application at hand?
D. If one or more applications already exists in the network and if yet another application is about to be developed, does the programming team know what tools and procedures are in use by the central support center for problem determination and system administration?

164. Programming Technique

Many application development organizations have found that a combination of structured design, pseudo-code, programming standards, and development inspections materially improves the productivity of their development teams (see Bibliography). Many of these techniques apply even if the programming team is as small as one or two individuals. These global techniques can be supplemented by a series of minor techniques that further increase programming quality and productivity:

A. Before programming, study the transaction frequency counts and the transaction sequence matrices developed during the analysis process. Then program so the most frequent cases are favored by the shortest path length.
B. While programming, list any assumptions about input ranges or input combinations which are critical to the code you have written.
C. As the code is developed, describe the test cases which will exercise it properly during unit test.
D. As the code progresses, list any error conditions that will be caught during processing.
E. Any time an I/O request is made, annotate that request with an informal priority that indicates whether the I/O is necessary to support a man-machine dialog, necessary to continue processing, or is needed to post all of the results

after processing is completed. Given that each I/O request carries a priority annotation, mechanizing a priority I/O queue will be relatively easy if one is needed to maintain performance.

165. Style Notes

The professional programmer is probably aware of many books on programming technique and style (see Bibliography). Most of these hints apply equally well to small, medium, and large processors. A few that seem particularly appropriate in the context of this handbook are:

A. For performance, try to block records to minimize I/O activity. Further, as soon as the key of the record desired is known, request the record, thus raising the probability that the record will be available before it is needed.

In addition, if the transaction sequence analysis shows a strong probability that a sequence of records will be required, call for the sequence as soon as the keys are known.

B. Where values of parameters are used to control a process (input edit limits, lists of legal values, etc.), organize these limits into a table and use this table to control what is accepted/transmitted. Then changes to limits can be made without changing the code.

In building tables, carry table length as an explicit parameter to be loaded prior to each table search. Treat the length of the table symbolically, using this parameter. Then new codes can be easily added.

Use this feature during unit test to compress the table to three or four entries for testing and then restore the full table prior to systems integration.

C. Program symbolically so the length of data fields, the sequence of data fields, and the population of data fields can be changed by a simple recompile.

D. On long reports keep a short push-down list of page breaks so printing can be restarted at the beginning of the sheet in the event of printer malfunction/operator inattention.

E. One programming team enumerated some rules for improving the performance of a data base application running on a large host machine. These rules are also appropriate when programming on a small machine.

1. The reference pattern of the application should touch the fewest concurrent pages during execution.

2. The mainline execution should be as straight a line as possible. The ideal program executes sequentially with no branch logic referencing beyond a small range of address spaces.

3. Literals and subroutines should be coded as close to their point of use as possible. (Duplicate the subroutines if necessary to avoid unnecessary page fetches.)

4. Exception routines should be isolated from the main code.

5. Do not alter code within a module. Prepare the code so it is reentrant. Then the code need not be paged out and perhaps the code will be reused without refetching.

DOCUMENTATION

166. Documentation Hints

Good documentation is usually produced by a specialist in technical writing. However, since distributed systems reach out into an organization farther than centralized systems, documentation for distributed systems users and their management needs to be prepared with special care. Even if the programming team has the services of a technical writer specializing in data processing, some additional familiarization and training may be required to acquaint him with the new audiences of users. He should get out into the organization and meet some of the intended users and their management.

He may wish to write some sample copy and ask actual users to review it for clarity. Then when given the list of documents, the tables of contents of each, and the estimated number of pages for each section, he will be able to understand the magnitude of his assignment. In addition, the technical writer should review the word processing and text editing tools available to verify their adequacy and should discuss with the programming manager what kinds of notes and draft copy will be prepared by the programmers.

If the programmers keep lists of input restrictions and error conditions, as suggested in the previous section, much of the detailed technical material will be captured for eventual publication in the formal documentation. If the designer has prepared a set of user scenarios, as was suggested in Chapter 4, the documentation specialist will have extensive material for his own self-education plus some useful user training material. Some specific hints to help the technical writer are:

A. Draw a form and prepare a series of one-page descriptions for all terminal commands. Be sure to list all error conditions for each.

B. If the system is going to provide an on-line HELP command, get the programmers to define how entries in the HELP library will be formatted and then prepare the on-line HELP text as soon as the technical details are provided by the programming crew.

C. Establish a set of notebooks in which background information about every command and report can be accumulated. Prepare an index to this set of notes so it is easily referenced (even during development) whenever the command, its mnemonic abbreviation, or its numeric code is known. Similarly index the reports by title, number, and the process producing the report. Cross reference all of these and publish a pocket card for use during development.

D. Be constantly alert for situations where two different transaction sequences can be used to accomplish the same result. Seek out information, from programmers and/or designers, so you know which is the preferred sequence given any set of conditions.

E. Note that the only truly new manual in a documentation series for distributed systems is the operations handbook for the local administrator. This should cover the environment, hardware, software, applications, support activities, problem determination, housekeeping, and site administration. Further, it must define the event logs and nonautomated administrative procedures he must set up to control security, assist in problem determination, and oversee the hardware, software, applications, and staff assigned to his location.

F. Depending on the situation, operators' procedures manuals, chief operators' manuals, and shift turnover checklists may be required. If remote sites involve large configurations and operate on two or more shifts to get all of the necessary batch processing accomplished, they will need procedures manuals and some instruction in process control techniques.

CONVERSION

167. Conversion Considerations

File conversion is not directly related to programming, although it is a major effort that occurs in the Programming phase. Conversion efforts range from moderate to large. Even if an application builds its necessary files during the processing of input transactions, it will probably still need tables of allowable values, rates, names and addresses, phone numbers of responsible individuals, and the like. These are major files (sometimes held as control tables) yet each one must be carefully developed and accurately converted for machine usage.

If the application under development replaces an existing automated application, file conversion may appear to be as simple as migrating an existing file onto the new system. Before leaping to that conclusion, the designer should obtain a program which edits the standing master files according to the edit rules to be implanted in the new program. One frequently finds that some data elements are altogether missing from the present files, some data elements are defined but are infrequently populated by values, and some data elements are uniformly present but with a quality which will be rejected when the new edit rules are applied. If any of these cases occur, then file work will be required which involves both clerical and programming personnel.

Frequently new distributed systems will require new automated files to be established from existing manual ones. For example, a parts wholesaler recently decided to automate his existing inventory control system. He had several hundred thousand manual stock record cards describing the parts in his five warehouses. The programming for the inventory system was almost trivial and was accomplished in 90 days by two persons. The problems relating to readying the manual files for keyboarding, getting them keyboarded, and accumulating changes to the newly keyboarded records until the new inventory system was in production occupied one full-time clerk and one part-time programmer for a year.

The installation of a broad-based manufacturing information system was recently reported in a popular magazine. Seven major applications were installed over a period of three years. The effort to establish clean vendor files, parts description files, and product structure files took half as much effort as the programming itself.

Each file conversion situation is different. Therefore no set formula or standard estimates reasonably apply. However, several situations reoccur with sufficient frequency to warrant note:

A. The data elements and quality of the new system may be more demanding than the existing system.

B. Manual files may contain known errors which do not inhibit manual processing, but must be cleaned up before the automated system can take over.

C. The manual files are not static, but continue to change after they have been automated. These changes must be recorded so they can be retrospectively entered into the on-line system when it becomes operational.

D. Cutover seldom occurs in an instant. More likely it is stretched out over a few weeks. The work in process (tasks initiated but not completed during the cutover period) sometimes requires special programming to be properly recorded in the automated files.

E. Some manual systems have extensive batch controls and audit trails to maintain processing integrity. The auditors are severely disappointed if these audit trails are not maintained during the process of conversion.

6
TEST PHASE

Figure 6.1 combines the task phasing and the activity list for the Test Phase. The principal active tasks during this phase are Integration Testing, Installation, and Initial Operation.

All the design hints given in this chapter may not apply in every case, but even a hint that does not apply may suggest a previously unrecognized area for study or investigation. It is suggested that each reader have a note pad handy as items unique to his environment may occur to him while reading the hints provided. If all personnel at an installation pool their notes, they will have augmented this handbook with items specific to their environment and class of work. Thus designers that follow will benefit from those who have gone before.

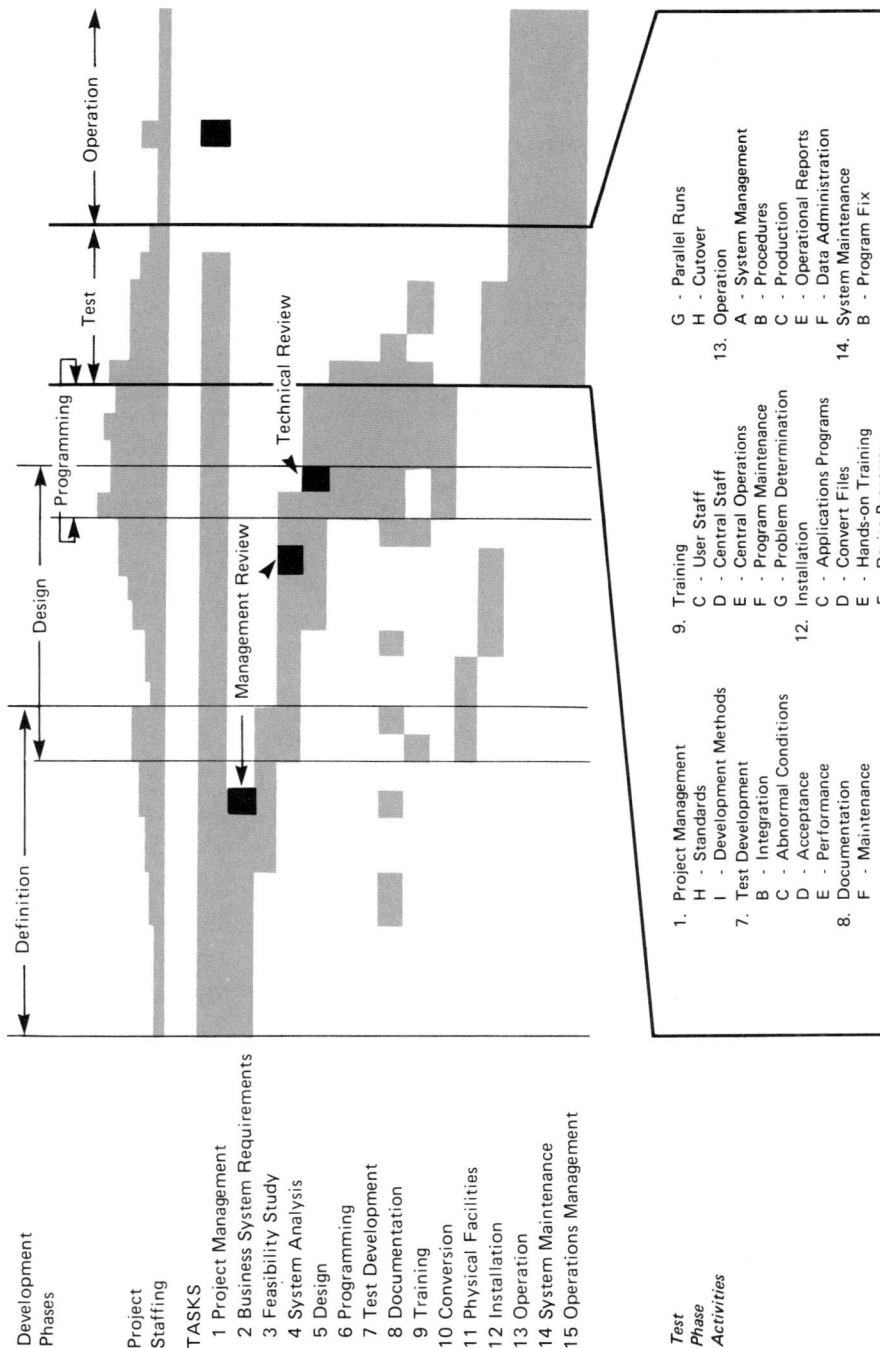

FIGURE 6.1 Test Phase

PROJECT MANAGEMENT

168. Management of Testing

Modern development technique brings to the Test Phase program modules which have been unit tested and test cases which have survived review by independent inspectors. In theory the programs will be integrated and the test cases exercised to prove that the development was properly conducted.

If a separate team of individuals was assigned the responsiblity for system test, and if they prepared system tests based on the design specifications, and if these test cases were then inspected (see Bibliography for technique), then the test case library will be in readiness for systems integration. The management challenge during the Test Phase is to:

A. Keep track of the program modules that have been tested and the test cases that have been run against them.

B. Verify the accuracy of the results produced by the running of each test.

C. Selectively rerun tests when errors are corrected in either the code or the test cases.

D. Be prepared to rerun all of the test cases as a unit against the final programs as a regression test step before the programs are released for initial installation.

E. Keep accurate records of all program changes made during testing activities so it can be verified that these changes were actually reflected in the source code, in the maintenance documentation, in the user manuals, and in the training materials.

169. Test Technique

If a large or complex system is being tested, a bookkeeping technique might be adopted from large systems experience. This technique causes temporary counters to be inserted at the branch points of the module under test so a record is maintained each time they are executed. If these counters are set to zero as the program is loaded, and count one every time the counter code is executed, the pattern of counts will reflect the degree of testing accomplished by one or more test cases. If the counter code is inserted in a program prior to compilation, it will be instrumented to report on test thoroughness. This code can then be removed before the program is released for production.

The counts so developed will allow the calculation of path length and provide transaction execution time estimates.

TEST DEVELOPMENT

170. Test Coverage

Test cases based on design specifications seldom test all of the error conditions recognized by the programming. If programmers kept separate notes on error conditions as they were coding (as suggested in Chapter 5), the test director can review these notes, see how many of the conditions are exercised by the tests already in the library, and prepare additional tests to cover the omitted conditions.

If training materials and user manuals are available in draft form, try to get some users on the system early to get a feeling for human factors and performance, items not easily covered when functional test cases are developed.

171. Test Feedback

Keep careful records of the situations encountered during integration testing to determine how many are likely to reoccur during production operation. Then check to be sure that the variety of situations addressed by the restart programs encompass those likely to reoccur.

172. Performance Cautions

If performance troubles occur, do not be in a rush to measure. Rather, methodically design a measurement experiment before gathering data. A performance experiment should cover assumptions, environment, workload, configuration of hardware and software, operator skill, analysis to be performed, what to measure, units of measure, uncertainty in operations, and conditions of observation (time of day, day of week, random workloads, selected sample, etc.). Beware of quick-and-dirty stopwatch timings as these can seriously mislead the development team. Also carefully sift reports of substandard throughput to determine if the system was being observed under production conditions and if all production procedures were being followed.

173. Test Reality

When devising and using load tests, remember the tests must be true to life. Consider operator training, mix of transactions, sequence of transactions, size of data base, and what other background activity may (should) be going on.

DOCUMENTATION

174. Documentation Tests

The system documentation must be tested along with the system. Documentation drafts can be given to users for review and in the later stages of integration testing, users can be allowed on the system to demonstrate their knowledge. Sometimes the indexes to the user documentation are deferred and hence are not available during testing. This should be avoided if possible as the indexes are extremely valuable during the testing period when time and patience are in short supply.

175. Documentation Controls

The maintenance of documentation, just like the maintenance of programs, starts during integration testing. Control procedures must be set up so documentation errors and ambiguous sections are noted whenever they are caught and corrected as soon as the pace of events allows.

TRAINING

176. Test Training

The training materials should be tested as part of the Test Phase. After the system operates more or less as designed, draft training materials can be used to acquaint a group of managers and a group of true-to-life users with the system. These two groups should probably be trained separately since some of the managers will have participated in the development and would not be satisfied with the pace required by neophytes. Yet some novice training should be conducted so the tone, content, and rate of delivery of the training material can be adequately judged.

177. Support Staff Training

If this application is the first one on the network, a special training course (probably coach and pupil sessions) will be required to educate the central support staff. Ideally one of the designers and one of the programmers would be transferred to the support staff for at least a period of six months or until the system settled down. If more personnel were required, then special courses will be needed to acquaint them with the philosophy, the concepts, the application, and the tools, and to assist them in setting up the support center procedures that are required.

    The earlier that the support center staff is put in place the better, so they can get experience in problem determination and repair while the original development team is still intact.

INSTALLATION

178. Phased Cutover

If the hardware has been installed and initial acceptance testing has been completed, the installation of a new system can take place in easy increments. Some combination of the following sequence may apply in almost every case:

A. Installation of the vendor's terminal emulation package so users at the remote terminals can communicate directly with the host.

B. Installation of a data capture program which allows downline load of control tables and code from the host, so local keyboarding can be locally edited and queued and then transmitted to the host at the end of the day.

C. Installation of a full-fledged application package (which may require temporary use of a tape drive for file loading) so local processing can be performed in the presence of local data files with summary information being supplied to upstream nodes.

D. Full installation of a complete distributed application which allows two or more nodes to communicate so files and data can be interchanged upon request.

All of these stages require application programs to be thoroughly tested before installation, files to be clean and converted, demonstration programs and hands-on training to be available for users, the manual processes external to the computer to be revised or established, parallel runs to assure the user about the reliability and integrity of the system, and a final decision to cutover so the previous processes can be discontinued.

OPERATION

179. Initial Operation

The previous chapters of the handbook have contained many items describing system features which will relieve some of the problems usually experienced during the initial operation of a new system. If the test code was prepared so that all terminal and file activity can be selectively recorded, turn it on. If programs were prepared to optionally zero out memory to eliminate the possibility that residue from one transaction can be used in subsequent processing, select the option. If the logging feature allows all console user and communications activity to be logged, enable it. If manual procedures are in place to provide accurate counts of input transactions and if file maintenance activity keeps a count of the records on the master file, compare the two numbers frequently. If manual diaries, trouble logs, and activity reports have been designed, implement them now.

Do not let the development schedule pressures inhibit good operational practice. Prohibit programmers from making ad hoc changes without full documentation. Insist that the production tools which leave audit trails be used even during initial operation. To do less will encourage ad hoc fixes so the system will run satisfactorily with the development programmers present, but will suffer significantly when the regular operations staff tries to run it normally.

After a time the local site manager will know which statistics, diagnostics, and measures provide him with the most information; which utility programs and restart routines have hidden capabilities not obvious from the documentation; and how to request help so he learns as the problem is being fixed. A good site administrator will keep a diary of these lessons so they may be distilled and passed on to administrators at other sites.

A. The on-site administrator needs crisp definitions and sample forms to properly report computer related events, downtime, repair history, reconfigurations, file rebuild, reprocessing following a failure, etc. Otherwise statistics will not be comparable from site to site.

B. On-site administrators need a diary log to allow recording of human actions that are not machine sensible.

C. Even if a system is designed to operate in a highly secure environment and to utilize encrypted communication links, the system will install and settle down much faster if no classified data is entered and no encryption is installed until the system operation becomes routine. Many of the development and service personnel will not have the clearances or need-to-know. A system which is secured too soon merely denies the local site administrator help from some of his most competent sources.

D. If the system configuration employs a local data base under the control of the site manager, the person who is to assume data administration responsibilities should be trained and in place, with the data administration procedures set up, before initial operation commences. The administrative activities related to bulk file load, test, file repair in the face of program errors, and file restore are more frenzied during initial operation than at any other time during the system's life. If control of file integrity is lost during this initial period, reloading the files in their entirety may be necessary before the system can be used for routine production.

SYSTEM MAINTENANCE

180. Maintenance Controls

If the suggestions given earlier in this handbook are followed, program fix procedures will be routine. Each program will have a version number embedded in it and every version installed will change this number. Any utility programs that apply temporary fixes will increment a counter so it is always possible to know how many fixes have been applied on top of a given base version. The site administrator will keep a diary to record the fix applied, the problem it was intended to solve, who applied it, and when it was done. If a remote site is being monitored by a central support center, information from the site administrator's diary would be forwarded to the center using operator-to-operator communication facilities so their logs and records can be updated.

Since none of the vendor software is perfect when delivered, they all have a procedure for recording symptoms, diagnosing problems, and determining whether the problems are in the hardware, the software, the vendor documentation, the application programs, or the user training. Hardware problems are usually fixed on site and immediately. If the problems in the software or the documentation cannot be readily identified, the symptoms and problem residue are usually sent off somewhere for in-depth analysis. Eventually a fix will appear or a programming restriction will be broadcast.

Several vendors transmit symptoms to a central location electronically and return fixes the same way. Others use couriers or the postal service. A good site manager will record all these events in his diary and communicate them to the support center so the list of open problems can be increased or decreased whenever one is discovered or permanently resolved.

If these controls are not installed during initial operation, there is a risk that the exact configuration of the application system will get out of control. Under this case a complete reload from the central library may be required and some of the testing may have to be repeated.

7
OPERATION PHASE

Figure 7.1 combines the task phasing and the activity list for the Operation Phase. The principal active tasks during this phase are Operation, System Maintenance, and Operations Management.

All the design hints given in this chapter may not apply in every case. But even a hint that does not apply may suggest a previously unrecognized area for study or investigation. It is suggested that each reader have a note pad handy as items unique to his environment may occur to him while reading the hints provided. If all personnel at an installation pool their notes, they will have augmented this handbook with items specific to their environment and class of work. Thus designers who follow will benefit from those who have gone before.

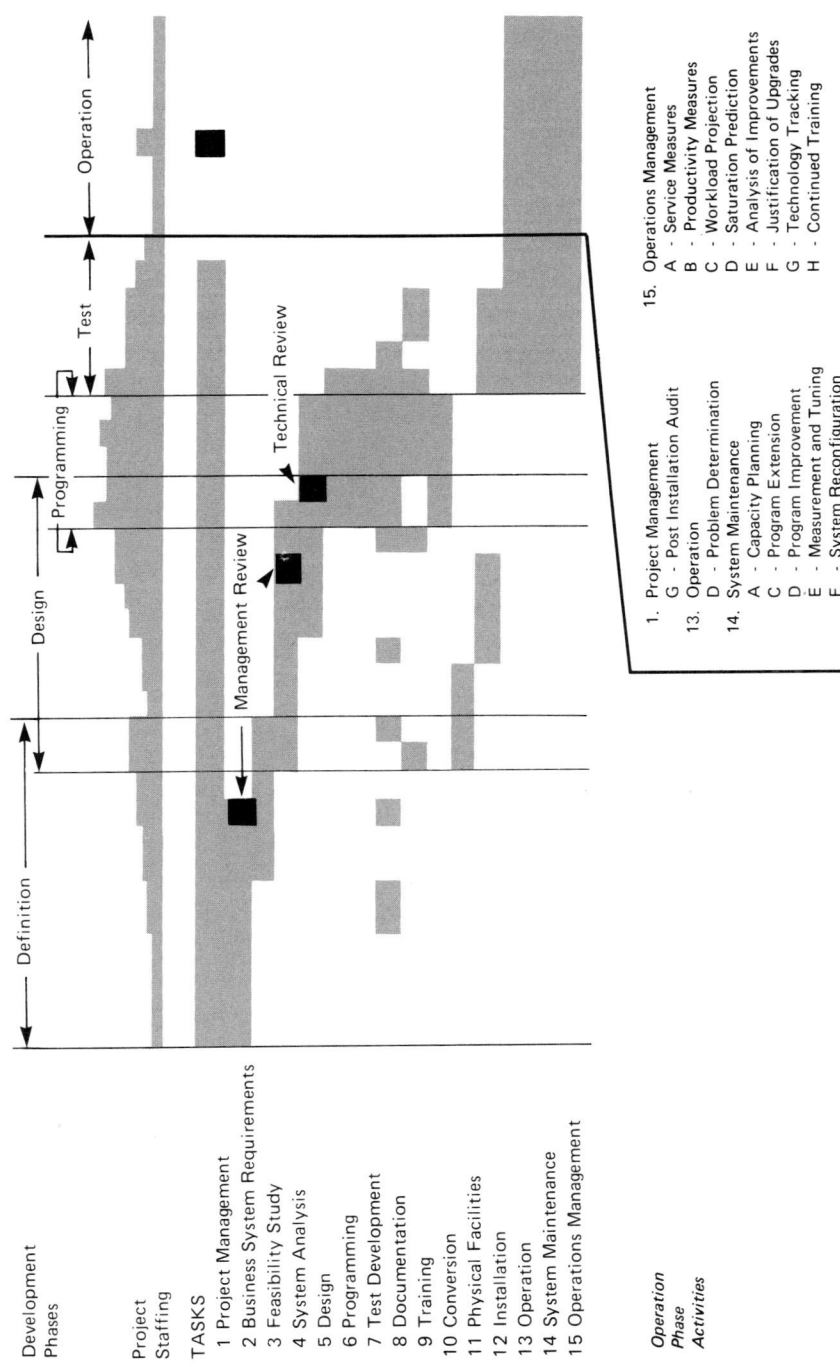

FIGURE 7.1 **Operation Phase**

PROJECT MANAGEMENT

181. Change Controls

The operators of a new system frequently feel lost and abandoned when the development crew disbands and leaves them on their own. Before the members of the development team get immersed in their next assignment, the operational crew must check to see if they have been adequately trained, if the documentation is current, and if they can live with the list of outstanding problems. In addition to coping with the day to day operational problems, they must review the notes and logs kept during integration test to determine which of the systems flaws have been corrected and which are pending. They should have corrected documentation, training materials, programs, and procedures, or schedule dates when these updated materials will be made available to them.

When corrections are received from the development team, the operational staff must decide how much testing to do on each correction and whether the fix is adequate for operational purposes. Prior to the official turnover for operation, the initiative lies with the development team while members of the operational staff assist or observe. Once turnover has occurred the operational staff carries the responsibility. The roles are now reversed and unless the operations crew asserts their needs vigorously, the required corrections may not be made.

182. Post-Installation Audit

The proper scheduling of the post-installation audit depends on the complexity of the system and the skills of the operational staff. While the operational crew is still learning at a rapid rate, things will not be routine. After a normal routine is established and each person has had an opportunity to use the training he received, and all major system flaws (if any) have been corrected, the site administrator should request an operational audit.

A knowledgeable, independent, objective person should lead the audit. Ideally this person would have had no part in the design, nor be responsible for any of the operation. The proper audit team would consist of one of the designers, one of the programmers, and a senior member of the operations staff. The goals of the project should be reviewed and each of the functions provided should be evaluated.

Each post-installation audit should conclude with an audit report. The contents of this report should cover the features planned versus the features delivered, the original schedule versus the final schedule, the original development cost estimate versus the actual cost incurred, and conclude with an enumeration of problems that still require correction. If the requirements changed during the development process or if the live environment and its staff differed from the planned environment, or if the project was reestimated and rescheduled one or more times during the development period, these matters should be covered in an explanatory appendix.

The purpose of a post-installation audit is to learn from your strengths so they can be repeated and to document your mistakes so they can be avoided. Some development organizations are not self-confident enough to stand a critical review and evaluation (even if it is conducted objectively by an independent party). In these cases dispense with the post-installation audit as it is likely to be counter-productive.

OPERATION

183. Problem Determination

The principal operation problem not adequately covered in previous sections of this handbook involves problem determination. System failures come in two varieties: hard and intermittent. Hard failures are relatively easy to diagnose and repair because the system responds in exactly the same way to exactly the same input. A hard failure may be due to a hardware malfunction, a bug in the software, or a flaw in the application program. But the important thing about hard failures is their consistency which makes them easy to locate and repair.

Some intermittent failures are due to an intermittent circuit in the hardware that may be sensitive to temperature, vibration, or static electricity. These are harder to locate since it is not easy to re-create the condition of failure. Intermittent failures in software and application code are usually due to flaws in the code which makes it sensitive to the lengths of queues, the instantaneous workload on the system, the sequence of transactions presented, or in some cases the values of numbers used in those transactions. Symptoms of intermittent failures are occasional wrong answers, unusual variability in execution time, and differing responses to consistent inputs.

If an intermittent failure is suspected, turn the logging feature back on so that it records the terminal dialog, all file activities, all commands from the console operator, and all messages from the communications network. There is a risk that when this is done, the internal processing pattern is sufficiently modified so that the intermittent failure does not reappear. If that occurs, then you have a new symptom and senior programming help should be called.

While logging all activities and looking for an intermittent problem, be sure to record all events that occur near the computer. Janitors running floor polishers, coffee pots in computer rooms (not recommended), maintenance of building air conditioning, and electric power fluctuations have all caused their share of intermittent problems. In chasing an intermittent, a full environmental log frequently provides the symptoms which lead to discovery of the problem and its solution.

184. Site Administration

The previous sections of this handbook have mentioned many features which will aid a site administrator in carrying out his operational responsibilities. Some of those bear repeating in this context:

A. Some system designs require periodic actions on the part of the systems administrator. He must print out operational statistics before the queue gets full, he must run a file reorganization utility to keep data space from becoming too fragmented, he must apply and test (at some mutually convenient time) the modifications and changes that have been sent to him, and if the system has experienced a failure and is operating on an abnormal configuration, he must restore the original configuration.

If the system is properly designed, documentation and checklists will remind the administrator of these actions, and manual procedures will direct him in accomplishing them. Each action will leave a telltale residue in storage somewhere that is interrogatable by the central support center so they can remotely read out statistics, determine the date and time the files were last reorganized, etc.

B. The programs used for system demonstration and the minifiles used for system test should be retained throughout the production life of the system so any user or administrator at any time can rerun the demonstrations for the benefit of guests or to gain assurance that the system is operating properly. In the event intermittent problems are encountered, operators may be unsure of the system and if they make a normal operating error, may be sufficiently suspicious to be unable to determine whether the error was caused by the system or their own actions. Demonstrations or other exerciser programs will allow operator confidence to be quickly rebuilt so full production can be reestablished.

C. If the system maintains activity statistics and if the designers worried sufficiently about the management needs of the site administrator for controlling his clerical workforce, optional reports will be available so statistics can be printed out for each job, terminal, and person. The production statistics will provide a measure of the workload, and error statistics will provide symptoms about system stability, operator proficiency, and the need for additional operator training.

D. It is not unusual for transaction processing time to change as the number of records in the data base increases. However, this change should gradually get less pronounced and flatten out as the data base reaches its designed size. Transaction count and machine loading statistics will help the administrator to track this phenomenon. If the measured response time continues to deteriorate with load, a programmer-analyst should be called.

SYSTEM MAINTENANCE

185. System Maintenance

All the changes made to a system after the development team has retired come under the heading of System Maintenance. That title covers capacity planning, program extensions, program improvements, measurement and tuning, and system reconfiguration. This handbook carries a consistent theme: plan for change and save expense. If this advice were taken, there is a plan for change and the documentation contains notes which tell programmers how to exploit the growth and flexibility that were provided whenever edit limits must be modified, additional reports added, or new applications accommodated. In addition, there may even be space reserved in message headers to identify new message types, to allow message compression algorithms to be used, or to route messages when additional nodes of the network are brought up.

In addition to all that has been said before, a few additional ideas help in the system maintenance process:

A. If a central library of source programs is maintained, it should be possible to set a flag whenever a programmer requests a copy of a program with the intent of modifying it. The librarian should also record whenever a modification is completed, approved, and cataloged. Then if two separate independent programmers unknowingly request the same program with the intent to modify it, the second requester can be alerted that there is a "change in process."

B. When reviewing transaction loads and error statistics, note if any one line, terminal, or operator has an abnormal amount of errors billed to them. If this appears to be true, and if one of the optional training functions allows a second terminal to display the dialog occurring at some other terminal, request a shift supervisor to enable the parallel display to determine if the error counts are legitimate. If they are accurate, determine how to provide additional training to those persons who require it.

C. Despite everything that has been said, the second application installed on a multipurpose small machine is likely to interfere with the first application. Obviously functional interference cannot be tolerated and must be repaired regardless of whether one or both of the applications require modification.

If the first application provided better terminal response in the absence of the second application, the users will see a change when the second application is brought up. If this change is significant, it is probably due to a design flaw in the first application. Before getting too concerned about the change or its magnitude, the performance specifications on the first application should be exhumed to see if the system is still performing within the specified limits even though the second application may have reduced the response time for the first.

D. The second application on the system may sufficiently impact the first so that tuning is required. This tuning may take several flavors.

If the two applications allow users to access the same data base, investigate changes to the external procedures so data entry is scheduled when the volumes of data inquiry are low (early morning, midday, and late afternoon in the case of an insurance branch office).

If the two applications are seeking different data bases, ask a programmer to review the allocation of data base space to the available disk files. While the programmer is in attendance, review the transaction frequency counts with him to see if there is an alternative allocation of computer memory to both programs and their data that will reduce the file access interference.

If a performance test tool was prepared during development which ran at a low priority and counted the resources available after the needs of production programs were satisfied, ask the programmer to reinstall it and measure the residual capacity of the system. Barring that, work with a programmer to estimate the capacity used by the measured transaction load knowing the path lengths and the I/O counts. It may be that the two applications running concurrently have saturated the processor or the disk and the only solutions are to reduce the load or increase the capacity of the hardware.

E. When the system was originally installed, it may have contained some conversion features to provide compatibility with the previous system, to clean up old files, or to assist operators in gaining familiarity with the new system. If sufficient time has passed and these conversion aids are no longer required, investigate removing them so the resources they use may be devoted to productive purposes.

OPERATIONS MANAGEMENT

186. Operations Management

The operational management responsibilities to be addressed are:

- service measures
- productivity measures
- workload projection
- saturation prediction
- analysis of improvements
- justification of upgrades
- technology tracking
- continued training

Depending on how the management of the network is organized, these responsibilities may be shared between the central and the remote locations. If the remote sites are mostly self-sufficient with their own programming, service, and administration, then the majority of these responsibilities will fall to the administrator at the remote site. At the opposite extreme, if the duties of the remote site administrator are assigned as an "additional duty" to the manager of the largest work unit that site serves, and if the site has no programming or service capability, then the central support center carries the heavier load (although some residual load still remains for the remote site administrator). Many other organizational choices exist in between these two extremes.

If the remote processor is supporting a stable application that was just migrated outboard from some host system, then past history on the transaction and file growth will be available for workload projection, saturation prediction, and justification of upgrades. However, if the application is new and all remote site personnel are inexperienced in data processing, several attempts may be necessary before good planning data is obtained.

The novice data processing user will not have the time or background to track computing technology. He needs some help from the support center to be sure that he does not continue to use inferior methods after the technology has taken a step. To do so reduces his ability to compete because the costs of obsolescent technology will increase his average labor rate.

There are several specific items which bear on operations management:

A. A cost allocation scheme provides good financial discipline to each operating unit. A cost accountant familiar with DP can gather all of the operating costs into a single pool and when the average workload is known, devise an average fixed cost for each terminal installed and variable unit costs for transactions and reports. Naturally there are several other cost allocation schemes, but almost any scheme provides useful input to line managers and useful planning factors for management.

B. If a system has terminals which are highly dispersed so the site administrator never has the opportunity to meet his users face to face, the capability for users to enter free-form messages at their displays and send them to the site administrator provides useful feedback on system operations. Further, text messages tend to be more positive, even when trouble is being discussed, than are telephone calls. Finally, hard copies of the text messages provide useful verbatim comments to systems designers contemplating improvements.

If the system has a broadcast feature, the site administrator can use this capability to send responses to his terminal users.

C. Several times this handbook has mentioned human factors, machine response, and keyboard rhythms. Even if the design was correctly done in the first place, each site administrator must monitor these important attributes to be sure that service levels do not deteriorate.

D. Each site administrator must keep track of the global work plans of his users. Then when major changes, upgrades, or additional applications are discussed, he can make sure that these changes are scheduled so they do not impact any local production activity peaks.

E. Designers and programmers make assumptions about the system's capacity, the user's tolerance for delay, error rates, etc. If history proves that some of these assumptions are grossly in error, the site administrator is likely to be the first to know. It is vital that he finds some constructive way to feed this back to the development teams so they can change their procedures.

F. Capacity planning has been discussed several times. Each site administrator must be aware of the delays he is likely to incur from the time he senses a need for increased capacity, to the time that that capacity is justified and an order is placed and then the new equipment arrives and finally is made ready for production use. Even if the vendor can supply instant hardware delivery, the other delays may total several months and include revised budgets, changes to the telephone service, tuning of the software, and modifications to the application programs to exploit the new equipment. The site administrator must know the likely total of these delays as this determines how far he must foresee saturation if his users are not to be impacted.

8 CONCLUSION

This handbook is aimed primarily at the system designer experienced in developing applications in a large host environment who is about to undertake his first application in a distributed environment. In attempting to show how the on-line distributed environment differs from the traditional centralized batch environment, several themes occur:

- Plan for change and save expense.
- Design is an iterative process.
- The user organization must change if it is to administrate and manage a distributed system.

It is intended that this handbook be read by every designer before his development project is initiated. While reading, the designer should note sections which are particularly appropriate to his application in his environment and should investigate any subjects which come up in the process of reading that are not adequately covered in this handbook. If design notes are available from other projects, distill their experiences before starting a new development effort.

This handbook leads designers through the five major application development phases: Definition, Design, Programming, Test, and Operation. Better systems should result from this tour since the ideas presented will cause each designer to think through the entire set of development phases before the project is initiated. As a result, the project plans and the work conducted should benefit materially. The following were highlighted:

1. life-cycle economics
2. human factors concerns
3. distributed data base design
4. network administration and management
5. application performance considerations

Readers may wish to reflect on the observations that a single application may dominate a small system, and that friendly distributed systems require that the manual and automated processes be jointly designed, implemented, and monitored.

This handbook is aimed principally at the designer who is building a commercial system to be operated by his countrymen in a single national environment. If a design team was building a large network which spanned two or more countries, problems with transborder data flow, differing languages, and differing national attitudes and work habits would have to be addressed.

If designers have need for systems with extremely high availability, they should not depend on this handbook for the necessary information. Although high availability is mentioned at one or two points, the depth of treatment is not sufficient for a designer faced with an extremely high availability requirement.

Several parts of this manual refer to the problem of networks which contain mixtures of new and old equipment or mixed systems containing equipment from several manufacturers. Even if your new system contains some special serviceability features to support remote service and problem determination, each mixed system is a special case and requires special analysis and design. Further, mixed systems may require some extra programming to operate effectively in a remote environment.

This manual is an attempt to aid designers of applications for a new environment. The problems discussed are real and need to be addressed by each design team. The advice given is the best available at this time, although some hints have not been thoroughly time tested. If better solutions are locally available in any development shop, do not hesitate to use them in lieu of the ideas presented here.

Remember that design is an intellectual, iterative process. Good design is a creative accomplishment. Designs which work, retain flexibility for change in the right places, and are in harmony with their users and their environment are the result of hard-working, well seasoned analysts. Less experienced analysts can progress more rapidly if they exchange hints like those contained in this handbook.

APPENDICES

A
PAYOUT WORKSHEET

Each distributed application will benefit from a Financial Payout Worksheet. As soon as a work plan begins to emerge, the project leader will be able to create his first schedule, list his costs, and estimate his benefits. Then given a set of assumptions, a Worksheet can be prepared. It matters little that the final hardware is yet to be selected, since intense competition between vendors yields most equipment costs about the same initially. The important act is to create the Worksheet early and see if the project appears to be justified (knowing the estimates will improve and the schedule will be revised). Then when better information is available, it can be reflected in the Worksheet.

The following pages contain a sample set of assumptions and a companion Worksheet. The resulting numbers form the basis for Figures 1.3 and 1.4.

NOTES — FINANCIAL PAYOUT WORKSHEET

1 — Wage Rates (column 2):
    Manager $3,000/mo.
    Lead programmer/analyst $2,000/mo.
    Designer/programmer $1,500/mo.
    Programming Support $1,000/mo.
    Operations $1,000/mo.

2 — Capital Equipment and Depreciation (column 5):
(Does not include any packaged applications programs)

    A — **Equipment List:**   Purchase Price
    Processor, Memory,
    Disk, Communications
    Adaptors, Magnetic
    Tape, 6 CRT Displays
    Line Printer and
    Miscellaneous Cables.
    Total $90,610

B — Depreciation: Assume 5 year life, straight line calculation, with 10% salvage value.

$$\frac{\$90{,}610 \times .9}{60} = \$1{,}359.15/\text{month}$$

3 — On-going Maintenance and Service (column 5):
(Does not include vendor programming assistance, if any)

                                                 Maintenance Charge
    A — **Equipment List:**
    Same as Enumerated in
    Schedule 2A (above)
    Subtotal $598/mo

    B — **Software List:**   License Charge*
    Operating System, Data
    Management Subsystem, CRT
    Support Subsystem, COBOL
    Compiler and Libraries,
    SORT, RJE, and Remote
    Service Support Package.
    Subtotal $565/mo

      *License for initial system

    C — Total $1,163/mo.

NOTES—FINANCIAL PAYOUT WORKSHEET

4 — Budget for depreciation, maintenance, and service (column 5):

    Depreciation $1,359
    Maintenance and Service   1,163
    Total $2,522/mo.

    $2,522/mo $\div$ 4.333 = $582/wk.

5 — The costs are shown in constant dollars. The following costs and benefits were not considered in the calculations:

    A —Any cost for user participation, training, or operations.
    B —One-time cost for travel, schools, and testing.
    C —One-time cost of training courses and materials.
    D —One-time cost of test time in excess of allotment.
    E —One-time cost for taxes and shipping.
    F —One-time cost for vendor installation assistance.
    G —On-going costs for communication circuits.
    H —On-going cost for power, air conditioning, and space.
    I —On-going cost for insurance.
    J —On-going cost for taxes.
    K —On-going interest on funds invested.
    L —On-going savings on discontinued quipment.
    M —On-going savings from discontinued service bureau contracts.
    N —Indirect savings derived from automated functions (reduced cost of inventory, faster clearing of receivables, less scrap and rework due to higher quality of information, etc.).
    O —Profit on new business made possible by automated functions.

PROJECT NAME: Satellite Information System

APPLICATIONS DEVELOPMENT

Events	1 Project Lifetime (Weeks)	By Week				
		2 Labor Expense ($000)	3 Facilities Expense ($000)	4 Supplies Expense ($000)	5 Depreciation Maint. & Serv. ($000)	6 Total Expense (2,3,4,5)
Project Start	1	1630	0	0	0	1630
	2	1630				1630
	3	1630				1630
	4	1630				1630
	5	1880				1880
	6	1880				1880
	7	2500				2500
1st Management Review	8	2500				2500
	9	2750				2750
Facility Started	10	3250	0			3250
	11	3250	2500			5750
	12	3250	2500			5750
	13	1880	2500			4380
Equipment Installed	14	2000	2500	0		4500
	15	2380	1000	2000		5380
	16	3750	0	0		3750
2nd Management Review	17	4000			0	4000
	18	4750			582	5332
	19	4000				4582
Formal Technical Review	20	6500				7082
	21	6000				6582
	22	5630				6212
	23	5880				6462
	24	5130				5712
	25	4880				5462
	26	3000		0		3582
	27	3000		100		3682
1st Production Operation	28	3000		100		3682
	29	2630		100		3312
	30	1880		100		2562
	31	1000		100		1682
1st Saving Occurs	32	1000		100		1682
Post Installation Audit	33	1000		100		1682
	34	2630		100		3312
	35	1000		100		1682
	36	1000	0	100	582	1682

DATE OF ESTIMATES 6-4-79   Page 1 of 2

FINANCIAL PAYOUT WORKSHEET

By Period	By Week					By Period	Project Cum
7 Accumulated Expense	8 Labor Saving ($000)	9 Facilities Saving ($000)	10 Supplies Saving ($000)	11 Equipment Saving ($000)	12 Total Saving (8,9,10,11)	13 Accumulated Saving	14 Project Net (13 minus 7)
1630	0	0	0	0	0	0	-1630
3260							-3260
4890							-4890
6520							-6520
8400							-8400
10280							-10280
12780							-12780
15280							-15280
18030							-18030
21280							-21280
27030							-27030
32780							-32780
37160							-37160
41660							-41660
47040							-47040
50790							-50790
54790							-54790
60122							-60122
64704							-64704
71786							-71786
78368							-78368
84580							-84580
91042							-91042
96754							-96754
102216							-102216
105798							-105798
109480							-109480
113162							-113162
116474							-116474
119036							-119036
120718	0				0	0	-120718
122400	1000				1000	1000	-121400
124082	1000				1000	2000	-122082
127394	1000				1000	3000	-124394
129076	1000				1000	4000	-125076
130758	1000	0	0	0	1000	5000	-125758

PROJECT NAME *Satellite Information System*

APPLICATIONS DEVELOPMENT

Events	1 Project Lifetime (Weeks)	By Week				
		2 Labor Expense ($000)	3 Facilities Expense ($000)	4 Supplies Expense ($000)	5 Depreciation Maint. & Serv. ($000)	6 Total Expense (2,3,4,5)
	37	1000	0	100	582	1682
	38	1000		100		1682
	39	1000		100		1682
	40	1000		100		1682
	41	1000		100		1682
	42	1000		100		1682
	43	1000		100		1682
	44	1000		100		1682
Proposed Savings Achieved	45	1000		100		1682
	49	1000		100		1682
	50	1000		100		1682
	51	1000		100		1682
1st Anniversary	52	1000		125		1707
	63	1000				1707
	64	630				1337
	76					
	88					
	100					
2nd Anniversary	104					
	116					
	128					
	140					
	152					
3rd Anniversary	156					
Breakeven	163					
	168					
	180					
	192					
4th Anniversary	208					
	214					
	216					
	228					
	240					
Project Net = $100 K	249					
5th Anniversary	260	630	0	125	582	1337
Breakeven plus 2 years	267	630	0	125	582	1337

DATE OF ESTIMATES 6-4-79   Page 2 of 2

FINANCIAL PAYOUT WORKSHEET

By Period	By Week					By Period	Project Cum
7 Accumulated Expense	8 Labor Saving ($000)	9 Facilities Saving ($000)	10 Supplies Saving ($000)	11 Equipment Saving ($000)	12 Total Saving (8,9,10,11)	13 Accumulated Saving	14 Project Net (13 minus 7)
132440	1500	0	0	0	1500	6500	-125940
134122	1500				1500	8000	-126122
135804	1500				1500	9500	-126304
137486	1500				1500	11000	-126486
139168	1500				1500	12500	-126668
140850	1500				1500	14000	-126850
142532	1500				1500	15500	-127032
144214	1500				1500	17000	-127214
145896	2000				2000	19000	-126896
152624	2000				2000	27000	-125624
154306	2000				2000	29000	-125306
155988	2000				2000	31000	-124988
157695	2500				2500	33500	-124195
176472						61000	-115472
177809						63500	-114309
193853						93500	-100353
209897						123500	-86397
225941						153500	-72441
231289						163500	-67789
247333						193500	-53833
263377						223500	-39877
279421						253500	-25921
295465						283500	-11965
300813						293500	-7313
310172						311000	+828
316857						323500	+6643
332901						353500	+20599
348945						383500	+34555
370337						423500	+53163
378359						438500	+60141
381033						443500	+62467
397077						473500	+76423
413121						503500	+90379
425154						526000	+100846
439861	2500	0	0	0	2500	553500	+113639
449220	2500	0	0	0	2500	571000	+121780

B
DIFFERENCE LISTS

A difference list is an analytical tool to allow two computing environments to be contrasted. Lists may be used to compare dialects of the same language, or program environments including language, data management subsystem, and operating system.

A difference list consists of three component parts. The left column is an exhaustive enumeration of all facets of the environments being compared. Each adjacent column contains a description of a system being compared. Each attribute of this comparison must be accurately described: Does the system offer the feature enumerated in the left-most column? If the feature is offered, what is the breadth of the implementation and what are the resulting limitations?

Difference lists are arduous to prepare, but when completed are invaluable for establishing programming standards or preparing design guidelines to insure compatibility between the two systems. Sometimes the two systems are not close enough to allow any practical level of compatibility, and in this case the designer must identify the points of incompatibility (where he had to use an incompatible feature to get the job done or maintain satisfactory performance) so the points to change will be identified if later it is necessary to migrate the code from one environment to another.

The difference list that follows is republished courtesy of the IBM Corporation. It compares two different versions of a PL/I compiler. This three part listing contrasts the features and limitations implemented in two language processors.

Language feature	Optimizing compiler implementation	Checkout compiler implementation
Statements: CHECK NOCHECK FLOW NOFLOW PUT SNAP PUT FLOW PUT ALL HALT %CONTROL	Syntax-check only	Implemented
Options: ORDER REORDER TOTAL	Implemented	Syntax-check only
Built-in subroutines: PLICKPT PLIREST PLICANC	Implemented	Syntax-check only
PUT DATA statement and CHECK prefix specifying program control data	Names of variables only transmitted	Names and values of variables transmitted
PUT LIST statement specifying program control data	Invalid	Values of variables transmitted
Lengths of pointer and offset variables	4 bytes	With COMPATIBLE compiler option: 4 bytes With NOCOMPATIBLE compiler option: 16 bytes
Oncodes	Certain codes not implemented (See list in section H, "On-conditions").	All oncodes implemented

Differences Resulting From Differing Compiler Functions

254   Appendix B

Language feature	Optimizing compiler implementation	Checkout compiler implementation
Based variable in data-directed I/O and CHECK name-list	1 Must not be based on an offset variable. 2 Must not be a member of a structure containing the REFER option. 3 Must not be based on a pointer that is based, defined, or a parameter, or a member of an aggregate.	No corresponding rules
Defined variable in data-directed I/O and CHECK name-list	Must not be defined: 1 on a controlled variable. 2 on an array with one or more adjustable bounds. 3 with a POSITION attribute specifying other than a constant.	No corresponding rules
CHECK prefix specifying label of statement to which prefix is attached	CHECK raised for the label	CHECK not raised for the label
LIKE attribute specifying a minor structure that is contained in a major structure of which some other minor structure is declared with LIKE attribute	Not allowed	No corresponding rule
Area variable in an OFFSET attribute either in DECLARE statement or RETURNS attribute or option	Must be non-defined, unsubscripted, unqualified area name	No corresponding rule
Area variable in OFFSET attribute in parameter descriptor	Not allowed	No corresponding rule
Data-directed output of dimensioned structure	Output in row major order for each array	Output as interleaved arrays in row major order
Exponentiation	See section F, figure f.4d	

Differing Qualitative Restrictions

Language feature	Optimizing compiler implementation	Checkout compiler implementation
Aggregate argument to generic entry name	Dummy argument cannot be created	No corresponding rule
Parameter string length or area size specified as other than decimal integer constant	Length or size attribute assumed to match argument: dummy never created	Dummy created if length or size differs from argument
Attributes of entry argument and parameter differ in alignment only	No dummy argument	Dummy argument created
Pseudovariables: COMPLETION COMPLEX PRIORITY STRING	Not allowed as control variables for do-groups	No corresponding rule
UNDEFINEDFILE condition in OPEN statement specifying more than one file name	Raised once, after attempting to open every file	Raised at each attempt to open a file that is undefined
Standard default files SYSIN and SYSPRINT	No corresponding rule	Used by compiler. Must not be declared with attributes conflicting with compiler requirements. SYSPRINT always open, therefore no new page started for program's output
Locator conversion (offset to pointer and vice versa)	1 If offset is a structure member, or if it appears in a DO statement or multiple assignment, the associated area must be an unsubcripted, non-defined element variable. If the area is based, its locator must be an unsubscripted, non-based, non-defined pointer, and it must not be used to explicitly qualify the area in the offset declaration. 2 Locator conversion cannot be performed between argument and parameter: both must be either offset or pointer.	No corresponding rules

Differing Qualitative Restrictions (Continued)

Language feature	Optimizing compiler implementation	Checkout compiler implementation
Maximum number of blocks in one compilation	255	No corresponding rule
Maximum level of nesting of blocks	50	No corresponding rule
Maximum number of active on-units	49 in any block 254 in any compilation	No corresponding rule
Maximum level of nesting of DO and IF statements	49	No corresponding rule
Maximum level of nesting of select-groups	49	No corresponding rule
Maximum level of dependency in DECLARE statement	1, except that an adjustable bound may not depend on a defined variable whose base: 1. is a parameter, 2. is automatic with adjustable extents, or 3. has fixed subscripts	No corresponding rule
Maximum number of entries in list of constants in declaration of COBOL variable	125	No corresponding rule
Maximum level of locator qualification	Depends on storage available, but never less than 10	No corresponding rule
Maximum length of character-string picture data	Depends on storage available, but never less than 1023	32767
Maximum number of subscripted label variables or subscripted label prefixes in one block	400 maximum. The exact number depends on the main storage available to the compiler, and is one tenth of the spill file record size. The maximum of 400 is obtained when the size of main storage available to the compiler is greater than 80k bytes	No corresponding rule
Maximum number of active qualified temporary results	32 for statements that generate intermediate temporary results for BASED, dynamically DEFINED, or subscripted variables	No corresponding rule

Differing Quantitative Restrictions

C
BIBLIOGRAPHY

BUSINESS SYSTEMS REQUIREMENTS

Burnstine, Donald C. *Business Information Analysis and Integration Technique* (BIAIT). Petersburg, New York 12138: From Author, 1978.

Carlsen, Robert D., et al. *The Systems Analysis Workbook.* Englewood Cliffs, New Jersey 07632: Prentice-Hall, 1979.

Gane, Chris, et al. *Structured Systems Analysis:* Tools and Techniques. New York, New York 10019: Improved System Technologies, 1977.

Gilmour, Robert W. *Business Systems Handbook.* Englewood Cliffs, New Jersey 07632: Prentice-Hall, 1979.

Hartman, W., et al. *Management Information Systems Handbook.* New York, New York: McGraw-Hill Book Co., 1968.

IBM Corporation. *Business Systems Planning:* Information Systems Planning Guide, GE20-0527. White Plains, New York 10604: Data Processing Division, 1978.

Patrick, Robert L. *Information Flow Studies.* Northridge, California 91324: From Author, 1970.

Ross, Douglas T., et al. *Structured Analysis for Requirements Definition.* Second International Conference on Software Engineering, IEEE, 1976.

Taggart, W.M. Jr., et al. *A Survey of Information Requirements Analysis Techniques.* ACM Computing Surveys, Vol. 9, No. 4, Dec. 1977: 273-290.

Teichroew, Daniel, et al. *PSL/PSA: A Computer-aided Technique for Structured Documentation and Analysis of Information Processing Systems.* IEEE Transactions on Software Engineering, Jan. 1977, 41-48.

HUMAN FACTORS

Engel, Stephen E., et al. *Guidelines for Man/Display Interfaces,* TR 00.2720. Poughkeepsie, New York: IBM Corp., 1975.

Gilb, Tom, et al. *Humanized Input.* Cambridge, Massachusetts: Winthrop Publishers, 1977.

Martin, James. *Design of Man-computer Dialogs.* Englewood Cliffs, New Jersey 07632: Prentice-Hall, 1973.

COMMUNICATIONS

Doll, Dixon R. *Data Communications - Facilities, Networks, and System Design.* New York, New York: John Wiley & Sons, 1978.

Miscellaneous. *Executive Guide to Data Communications.* New York, New York 10020: McGraw-Hill Publications Co., 1978.

Sharma, R. L. *Integration of Advanced Communications Techniques.* Proceedings of the Terminal-based Systems Conference, AIIE. New York, New York, 1978, 563-586.

PROGRAMMING TECHNIQUE AND STYLE

De Marco, Tom. *Structured Analysis and System Specification.* New York, New York: Yourdon, 1978.

Gilb, Tom. *Data Engineering.* Lund, Sweden: Studentlitteratur ab, 1976.

Iwahashi, David S. *Order and Discipline: Benefits of Structured Techniques.* Datamation, October 1979: 150-166.

Jackson, Michael A. *Principles of Program Design.* London, New York: Academic Press, 1975.

Kernighan, Brian W., et al. *The Elements of Programming Style.* New York, New York: McGraw-Hill Book Co., 1974.

Myers, Glenford J. *Software Reliability* (Principles and Practices). New York, New York: John Wiley & Sons, 1976.

Peters, Lawrence J., et al. *Comparing Software Design Methodologies.* Datamation, November 1977: 89-94.

Rudkin, Ralph I., et al. *Structured Decomposition Diagram: A New Technique for System Analysis.* Datamation, October 1979: 130-146.

Stevens, W.P., et al. *Structured Design.* IBM Systems Journal, Vol. 13, No. 2, 1974, 115-139.

Weinberg, Gerald M. *The Psychology of Computer Programming.* New York, New York: Van Nostrand Reinhold Co., 1971.

Yourdon, Edward. *Techniques of Program Structure and Design.* Englewood Cliffs, New Jersey 07632: Prentice Hall, 1975

TEST INSPECTIONS

Crossman, Trevor D. *Some Experiences in the Use of Inspection Teams in Applications Development.* Proceedings of Application Development Symposium, IBM/SHARE/GUIDE, October 1979, 163-168.

Fagan, Michael E. *Design and Code Inspection to Reduce Errors in Program Development.* IBM Systems Journal, Vol. 15, No. 3, 1976: 182-211. (Reprint Order G321-5033)

Fagan, Michael E. *Inspecting Software Design and Code.* Datamation, October 1977: 133-144.

INDEX

Abnormal
  condition module, 109, 121
  conditions, 23, 63, 219
  environments, 107
  errors counts, 234
Aborting the system, 107-108
Acceptance, 23, 24, 63, 219
Access
  control(s), 24, 63, 174, 205
  inhibited, 178
  lists, 87
  privileges, 122
Accommodating communications error, 113
Accommodating error, 107
Activities check list, 21

Activity
  highlights, 26-32
  information, 130
  list-design, 63
  lists, 87
  log, 130
  patterns, 87, 90
  statistics, 95
Addresses into paths, 130
Addressing in a network, 112
Ad hoc changes, 225
Administrative
  change control system, 117
  controls, 42
  features, 125-127

Administrator's manual, 128
Air conditioning, 24, 33, 61
Allowable values, 215
Amending the estimates, 55-56
Amending the plan, 55
Analysis
   for distribution, 83-86
   matrices, 83-84
   matrix
      for screens, 141
      for training, 202
   of improvements 25, 229
   outline, 46-47
   program, 196
Analyst's value judgements, 47
Anticipating change, 116
Anticipating record calls, 212
Application
   economics, 3-17
   feasibility, 50-51
   overview, 200
   portability, 53-54, 103
Applications
   interference, 70
   pass-through, 127
   programs, 24, 219
   repair, 104
   specific functions, 42
Archival files, 92
Audit report, 231
Audit trail(s), 23, 30, 70, 116, 126, 155, 175, 176, 177, 217
Authorization control, 82
Authorized users, 123
Availability
   design, 187
   high, 106, 240
   maximization, 110
   standards, 72-73

Backing out transactions, 189
Backup and recovery, 106
Bad hookup, 115
Bandwidth analysis, 161
Bandwidths, 160, 163
BASIC, 41
Batch
   processing, 159
   update, 156
Benefit estimates, 99-101
Big ticket controls, 175

Blackout, 156, 157, 160, 178, 188
   period, 154
   tolerance, 111
Blanking buffers, 197
Blinkers, 132
Blocking
   factors, 155
   records, 212
Breakpoint messages, 128
Breaks in the data, 150
Broadcast
   feature, 237
   messages, 128
Built-in
   flexibility, 96
   statistics, 96
   training mode, 95
Business
   forecasts, 76
   level stability, 148
   system requirements, 6, 22, 26, 44-49
   systems planning guide, 26

Capacity
   estimates, 76
   loading, 51-52
   planning 25, 76, 122, 229, 238
Cash flow, 12-14
Casual operation checklist, 138
Casual terminal operation, 140-141
Cataloged search commands, 124
Cataloging processing modules, 90
Catastrophe, 154, 155
Catastrophe-if-wrong controls, 175
Catastrophe-if-wrong items, 66
Catastrophe recovery, 92
Catchup mode, 192
Central fix log, 172
Central (network) support center, 30, 71-72, 97, 104, 109, 113-115, 128-130, 172, 173, 177-178, 200, 202, 223, 226, 236
Central
   operations, 24, 219
   problem log, 172
   review group, 103
   service standards, 71-72
   staff, 24, 219
   support center. *See* Central network support center.
Centralized development standards, 103
Certification-for-release, 175

Index   263

Chained processing modules, 166
Change
  control(s), 22, 25, 41, 63, 230
  in process flag, 234
  in the business, 75
  level number, 176
Check digit checking, 176
Check digits, 65, 72
Checking modules reinvoked, 197
Checkpoint records, 155
Checksums, 154
Clock offsets, 130
Clocks and lockouts, 66
Closing the books, 91, 95
Coaching trainees, 180
COBOL, 41
Coded variables on reports, 138
Coding
  and abbreviation, 123
  hints, 213
  standards, 67
Cold start, 72-73, 74, 129
Collecting existing forms, 56
Collecting reference lists, 57
Combining files, 151
Command definition index, 214
Communication
  functions, 42
  specialist, 158
Communications, 23, 24, 63
  administration, 130-131
  and network aspects, 23, 63
  aspects, 23
  capacity, 59
  line loading, 50
  management, 165
  patch panels, 131
  protocols, 161-164
  traffic analysis, 58-59
  troubles, 113
Compatibility, 112, 163, 252
Completeness flag, 154
Complexity
  network, 19
  system, 19-20
Compression flag, 163
Configuration, 22, 63
Configuring for peak load, 122
Consistency
  on CRT's, 135
  screen, 69

Consistent screen display, 134
Console controls, 176
Context editor, 170
Continued training, 25, 229
Continuing processing, 190
Control tables, 65
Controlled shutdown, 108
Controls, 23, 63, 173
  and audit trails, 24, 208
  in network, 116
  system, 28
Conversion, 6, 23, 24, 28, 30, 58, 63, 203-205, 215-217
  considerations, 215-217
  design, 63
  programming, 208
  strategy, 203-205
Convert files, 24, 219
Converting existing files, 216
Coordinated access, 119
Cost
  allocation, 237
  benefit summary, 101-102
  payoffs, 101
  tracking, 38
  tradeoffs, 47-48
  re-estimating, 97
Counters inserted in tests, 221
Counts for controls, 175
Coverage programmers, 71, 104, 177
CPU load estimating, 51
Critical response times, 133
Critical skills, 39
Cross reference lists, 86
CRT standards, 68
Cutover, 25, 219
Cutover phases, 223

Data
  administration, 25, 219
  bank, 23, 63
  base
    activity log, 197
    architecture, 147-150
    definition, 147
    design, 87-93, 152-154
    distribution, 150-152
    loading, 155
    management functions, 42
    management system, 156
    performance, 155

Data (con't.)
  base
    redundancy, 149
    splits, 149
  capture
    module, 44
    priority, 135
    program, 224
  cross reference lists, 86
  description, 23, 208
  dictionary(s), 86, 157-158
  distribution, 148
  edit module, 44
  gathering, 22, 34
  log file, 44
  logging selection, 178
  quality improvement, 101
  seeks per transaction, 181
  set, 147
Dating files, 153, 175
Debug standards, 70-71
Debugging
  mode, 125
  production, 197
  tools, 71
Decentralized maintenance, 103
Decomposition, 23, 29, 63, 193
Deferred transactions, 174
Definition phase, 7, 33-61
Definitions of basic terms, 147
Degraded operation, 190-192
Degraded service, 188
Delay tolerance, 182
Delayed updating, 120
Deleted records, 67
Deliverable items, 37
Dependency chart, 35
DESCRIBE command, 140
Design, 6, 22, 27, 102-193
  authorization, 48
  classifications, 42
  decomposition, 193-195
  documentation, 86-87
  flexibility, 96
  for performance, 180-182
  for testing, 184-185, 197-198
  hints, 62
  phase, 7, 62-206
  separation, 57
  team training, 202
Designing for change, 117-120

Development
  activities, 20-25
  and operations cost targets, 22, 34
  inspections, 211
  methods, 22, 34, 63, 208, 219
  standards, 172. *See also* Standards and Programming standards
  task phasing, 6-10
  team, 98
Device allocation tables, 109
Device independent programs, 195
Diagnostics categorized, 177
Diagnostic messages, 116
Dialog
  designing, 133-134
  host/periphery, 59
  standards, 68-69
Dictionary
  data definitions, 86
  driven dump formatters, 196
  procedures, 158
Difference lists, 54, 99, 252-256
Disabling automatic-routing, 164
Disk
  allocations, 183
  capacity estimates, 50
  overloading, 93
  seek sequencing, 183
Display screen layout, 119
Distributed
  accountability, 175
  data, 52
  systems documentation, 86
Distribution lists, 170
Ditto capability, 143
Document
  assumptions, 60
  current processes, 22, 34
  special procedures, 58
Documentation, 6, 23, 30, 59-60, 200-201, 213-215, 222
  controls, 222
  design, 63
  hints, 213-215
  strategy, 200-201
  techniques, 59-60
  tests, 222
  updating, 104
Documenting the existing system, 57-58
Double update, 157
Downline program load, 177

Index

Downtime actions, 105
Downtime statistics, 45
Dummy reports, 43
Dump/restore, 155
Duplicate
   data base, 89
   messages, 70
   transaction sequences, 214

Edited field controls, 141
Editing, 117
Editing existing files, 215
Efficient coding hints, 213
Employee morale, 74
Emulation (in communications), 127, 224
Encryption devices, 131
End-of-day processes, 66
Environment, 22, 63
Environment of the system, 19
Environmental log, 232
Equipment, 24, 63
   units, 160
Error
   accommodation, 107-108
   checking, 72
   counts, 125
   logging module, 169
   logs, 130
Error-prone environments, 107
Estimating
   benefits, 99
   response, 132-133
   throughput, 93
Evening cleanup, 94
Event logs, 130
Exception processing, 150
Exclusive control of data base, 178
Exercise programs, 129
Existing system documentation, 57-58
Expected elapsed time, 200
External processing, 171-172
Extract program, 198

Failure analysis, 188
Failure without warning, 114
Feasibility study, 6, 22, 26, 50-56
Feedback from use, 238
FIFO order, 69
File
   access methods, 83
      privilege, 87

File (con't.)
   backup, 92
   catch-up, 91
   cleanup, 24, 208
   communications tradeoffs, 158-161
   conversion, 29, 30, 203, 209, 215, 216
   cutover, 216
   ID's, 176
   dumping, 92
   editors, 98
   integrity control, 226
   key, 147
   ranges, 87
   keys, 81, 87
   loading and recovery, 155-157
   rebuilding, 155
   recreation, 106
   security, 95
   size constraints, 50
   structure, 147, 149, 159
Fill-in-blanks, 135
   screen, 134
   mode, 180
Financial controls, 22, 34, 38
Financial payout worksheet, 12-15, 101, 245
Fix counters, 226
Fix logs and standards, 172
Flat file, 147
Flawed tooling, 198-199
Flexible designs, 109-110
Follow-on training, 179, 203
FORTRAN, 41
Frequency histogram, 75, 87
Full function system, 191
Functional goals and objectives, 22, 34
Functional interference, 235

Global process controls, 173-175
Global work plans, 238
Goals of distribution, 44-45
Growth, 60, 106

Hands-on training, 25, 219
Hard-copy communications, 126
Hardware diagnostics, 129
Headcount, 100
HELP
   capability, 144
   command, 179, 201, 214
   groups, 205
Hierarchial file structure, 147

High availability, 187-190
　configurations, 110
High volume activity, 94-95
HOLD status, 126
Host system, 28, 30, 78, 92, 132, 158
Human factors, 122-124, 131, 134, 238, 240
　briefing, 210
　hints, 134-137
Human intervention, 130

Immediate messages, 128
Incremental
　conversion, 185-187
　costs, 48
　installation, 204
Independent testing, 181
Infrequent activities, 94
Initial operation, 224-226
Input, 23, 208
　conventions, 64
　output matrix, 80
　ranges and combinations 211
Inspection of test cases, 220
Installation, 6, 24, 30, 76, 206, 223-224
　design, 63
　manual, 61
　non-standard, 51
　of equipment, 30
　planning checklist, 205
Installing change, 52-53
Installing incrementally, 185
Instruction path lengths, 181
Integration, 23, 63, 219
Interactive dialog, 159
Interactive systems, 135
Interference between applications, 181
Intermittent failures, 232
Internode messages, 112
Internode traffic, 90
Interview feedback, 47
Interviewing users, 45
Interviewing techniques, 45-46
Investment profile, 14-17
I/O overlap, 195

Justification of upgrades, 25, 229
Justifying the application, 99

Key range(s), 85, 87
Kickoff checklist, 210-211

Large file updating, 92
Learning curve, 96
Leontieff, V., 80
Library of data sets, 128
Library of programs, 43, 234
Life-cycle
　activities, 18-32
　approach, 5
　design, 178
Limited compatibility, 112
Line time vs. message length, 161
Linked transactions, 149
Lists in use by workers, 57
Living with error, 107-108
Locations lists, 87
Location-sensitive information, 65
Locked file sets, 189
Lock/unlock, 155
Log file, 67
Log outline 44
Logging
　message dialogs, 125
　module, 169
　program, 196

Maintaining documentation, 222
Maintenance, 24, 219
　controls, 226-227
Management
　metrics, 45
　of change. See Change.
　of testing, 220
　reporting, 152
　review, 10
Man-machine dialog, 22, 28, 63, 133, 200
　techniques, 141
Manual backup, 191
Manual procedures, 46, 56, 57, 124, 171, 173, 174, 191
Maximum blackout time, 111
Measurement
　and tuning, 25, 229
　experiment, design, 221
　of success, 45
Mental set of operator, 132
Merge, side effects, 151
Mesh network routing, 164
Message
　acknowledgement, 128
　categorization, 129
　descriptors, 70

Message (con't.)
  dialog logging, 125
  handling, 113
  ID, 67
  logs, 130
  routing, 112-113, 164-165
    information, 164
  standards, 69-70
  switching, 159
  tagging, 130
Migration, 96, 162
Milestone reporting, 22, 34, 40
Minibatching, 167
Minibatch processing, 191
Minimum configuration, 183
Missing data, 64
Missing input data, 136
Mixed character sets, 163
Mnemonics, 144-145
Modem switching, 113
Modes of data entry, 143
Modular structure, 44
Monitor module, 198
MORE command, 123
Morning startup, 94
Multinational corporation, 89
Multi-node networks, 127, 204
Multiple
  applications, 66, 70
  data sets, 128
  files, 119
  indexes, 151
  processes, 149
  screens, 145
  terminal sessions, 54

Name and address handling, 64
Native availability, 187
Neophyte training, 223
Network
  administration, 127-130, 172-173. See
    also Central network support center.
  characteristics, 115-116
  complexity, 19
  controls, 177-178
  management, 115-116
  PERT charts, 9
  problem management, 116
  support center. See also Central network
    support center.
  testing, 70

New product introduction, 76
Node data ownership, 88
Node identification, 152
Non-standard I/O, 51, 55
Non-stop operation, 106, 108, 121

One-to-one environment, 150
On-line benefits, 101
On-line update, 120
On-site
  administrator, 225, 226, 237-238
  hardware maintenance, 104
  service, 129
Open-ended tables, 170
Operation(s), 6, 25, 31, 224-226, 231-233
Operation phase, 7, 228-238
Operational
  audit, 22, 229, 230-231
  environment, 104
  management, 31
  reports, 25, 219
Operations handbook, 215
Operations management, 6, 25, 31, 236-238
Operator, 24, 208
  absent installation, 114
  absent operation, 108-109
  action recording, 176
  commands, 122, 123, 139
  comments table, 143
  confidence, 233
  intentions, 141
  options, 146
  procedure manuals, 215
  proficiency level, 123
  search options, 134
  support center communication, 97
Operator's manual, 94-95
Optimum paths, 164
Optional plans, 38-39
Orderly shutdown, 108
Orderly startup, 108
Organizational change, 47, 52-53, 55, 102,
  204, 239
Organizational stability, 148
Output, 23, 208
  checking, 66
Outside resources, 36-37
Overload, 69, 193
Overview document, 150

Packed records, 182
Page breaks restart, 212

Paper backup, 173
Parallel operation, 186
Parallel runs, 25, 219
Parameterized limits, 212
Parametric inputs, 196
Part-time console operators, 138-140
Passive log, 129
Pass-through code, 127
Path reliability, 165
Patterns of activity, 47
Payout worksheet, 12-15, 101, 245-251
Peak communications traffic, 58
Peak load configuring, 122
Peer review, 36
Percent complete reporting, 37
Performance. See Saturation.
Performance, 23, 28, 55, 63, 219
    cautions, 221-222
    degradation, 181
    estimating, 51
    measures, 183-184
    oriented designs, 180
    tooling, 199-200
    trade-offs, 182
    with growth, 183
Periodic actions, 232
Personal command lists, 124
Personnel evaluation, 39
PERT charts, 9
Phased cutover, 223-224
Physical
    facilities, 7, 24, 30, 60, 205-206
        design, 63
    installation, 206
    limitations, 50
    planning, 30-31
Pilot exercise, 98
Pilot installations, 204
Plan for change, 40-41, 116-117
Plan review, 22, 34
Planning for the future, 102
Polling inhibited, 193
Positional layouts, 142
Post-installation audit, 22, 229, 230-231
Power, 24, 34
    failure, 154
    requirements, 60
Preliminary
    acceptance, 24, 63
    cost justification, 22, 63
    design, 22, 47, 63
    estimates, 22, 34

Preliminary (con't.)
    operational costs, 48-49
    operator's manual, 94-95
    user's manual, 43
Prenumbered forms, 81
Primary file indices, 147
Printed output headings, 126
Printed reports ordering, 137-138
Priority
    codes, 69
    costs, 77
    I/O queue, 212
    levels, 135
    messages, 69
Privacy, 82
Private search commands, 124
Problem determination, 24, 25, 30, 72, 103, 115, 128, 219, 229, 231-232
    aids, 129
Problem history, 176
Problem prediction, 117
Procedures, 23, 25, 63, 219
Processing (design), 23, 63
Processing overload, 51
Processing (programming), 23, 208
Production, 25, 219
    program catalog, 177
    statistics, 233
Productivity measures, 25, 229
Program
    extension, 25, 229
    fix, 25, 219
    fix procedures, 226
    improvement, 25, 229
    maintenance, 24, 219
    release number, 177
Programmers, 24, 63
Programming, 6, 23, 29, 193-195, 208, 210-213
    design, 63
    for maintenance, 169
    kickoff, 210-211
    languages, 41
    notes, 165-171, 212, 214
    phase, 7, 207-217
    standards, 41, 64, 70, 116
    standards review, 210
    technique(s), 211-212
Progress indicators, 132
Progress tracking, 22, 34, 37, 210
Project
    assumptions, 36
    costing, 12, 97

Project (con't.)
  estimating, 38
  management, 6, 22, 26, 35-44, 64-73, 209-210, 220-221, 230
  design, 63
  plan, 22, 34
  planning, 35-36, 209-210
  staffing, 39
  standards, 41-42
  status, 37-40
Prompting, screen, 69
Proper names, 64
Protocol flag, 163
Proven search commands, 180
Pseudo code, 211

Quality assurance counters, 71
Queue limits, 109
Quick look capability, 146

Range lists, 215
Rate
  charting, 210
  of change, 204
    control, 53
  tables, 118-119
Read-only data base, 120
Rebuilding master records, 153-154
Reconfiguration, 117
Reconciliation of file data, 91
Reconstructing records, 153
Record structures, 147
Records definition, 147
Recovery, 91, 110, 111
  actions, 94
  logs, 157
  sequences, 189
Redundancy in data base, 148
Redundancy protection, 106
Redundant loops, 107
Reentrant code, 213
Reentrant programs, 183
Re-estimating costs, 97
Regression test system, 196
Regression testing, 198
Remote
  diagnosis, 104
  problem determination, 104
  testing, 185
Replanning the project, 209-210
Replicated data, 88
Replicated data base, 90

Replication protection, 106
Report mock-ups, 43
Reports
  design, 137-138
  formatting, 171
  location matrix, 84
  routing, 170
  standards, 67-68
  systematically collected, 56
Requirements, 23, 34, 48, 49, 52
Resource allocation, 71, 183
Response
  maximization, 88
  time, 109, 131
    constraints, 167
    scenarios, 133
    statistics, 45
Responsibility assignment, 230
Responsibility for remote location, 236
Restart, 91, 96, 110, 111, 153, 155, 157, 170, 189
  procedures, 108
  programs, 98
  recovery and availability, 23, 63, 193
Return on investment analysis, 12
Reverse index, 153
Reversing transactions, 149
Review programming standards, 210
Revise processes, 25, 219
Revised cost estimates, 97-99
Rework avoidance, 102
Risk analysis, 110-111
Roles and missions, 102-105
RPG, 41
Rules
  for design, 117-120
  for human factors, 122-124
  for man-machine dialogs, 134-135
  for non-stop operation, 121
  for reports design, 137-138
  for system administration, 125-127
Run time parameters, 96

Sample test files, 95
Saturated disk, 94
Saturation, 236
  indication, 135
  of the system, 131
  prediction, 25, 229
SAVE command, 125
Schedule constraints, 22, 34, 49
Scheduling suggestions, 49
Scherr, A. L., 112

Screen
   design, 141-146
   overflow, 145
   protocols, 69, 135-137
Scrolling screens, 134
Search options, 134
Search output headings, 138
Second application interactions, 235
Secondary index, 153
Secondary transactions, 163
Security, 155, 174, 189, 205, 225
   features, 131
   lists, 87
   requirements, 125
   risk, 113
Seek counts, 151
Selective logging, 125
Selective routing, 127
Self-defined headings, 126
Sequential transaction numbers, 81
Service measures, 25, 229
Service messages, 132
Sharma, Roshan Lal, 160
Shift turnover checklists, 215
Shutdown, 96
Simultaneous access, 174
Simultaneous transactions, 150
Single application saturation, 94
Single module dialog, 168
Site
   administration, 232-233
   manager's diary, 226
   manuals, 128
   planning, 205-206
SNA architecture, 162
SNA features, 162
Snyder, Terry, 210
Software
   errors, 226
   extensions, 42
   problems, 226
Sort step (nullified), 119
Space, 24, 34, 60-61
   requirements, 60
Special
   power lines, 61
   symbol printing, 68
   tools, 98
Specific design goals, 105-107
Specific process controls, 175-176
Specifications, 23, 63

Split
   data base, 90
   files, 88, 150-151
   processing, 167
   screen, 135
Staff loading, 10-12
   chart, 11
Staffing and talent, 22, 34
Staffing for installation planning, 9-10
Stages of processing, 194
Standards, 22, 34, 63, 208, 219. *See also*
   Development and Programming standards.
Standards for downtime reporting, 225
Standing file conversion, 24, 208
Startup
   characteristics, 76
   procedures, 108
   steps, 224
Statistics on transactions, 45
Statistics standards, 225
Stopwatch timing, 222
Storage size constraints, 166
Structured
   applications, 73
   design, 211
   programs, 119
Style notes, 212-213
Style of manuals, 200-201
Support center
   control, 128
   messages, 113-115
   features, 97
Support staff training, 202, 223
Survivable applications, 190
Synchronized data bases, 126
Synchronizing clocks, 66
System
   administration, 125
   aspects, 22, 63
   complexity, 18
   crash, 96
   decomposition, 117
   failures, 231
   life-cycle, 5
      activities chart, 22-25
   maintenance, 6, 25, 31, 226-227, 234-236
   management, 25, 219
   overview, 24, 63
   reconfiguration, 25, 229
   saturation, 180
   status indicator, 190

System (con't.)
   test programs, 232
Systems analysis, 6, 22, 27, 56-59, 73-102
   design, 63
   overview, 30

Table driven, 118
   code, 93
   error module, 169
   test data, 196
   test tools, 71
Table length paramenter, 212
Table of limits, 66
Tape labels list, 140
Task
   chart, 35
   enumeration, 35
   phasing, 63
      chart, 7
Technical writing suggestions, 214-215
Technology tracking, 25, 229, 237
Technology vs. requirements, 22, 34
Telephone interface, 158
Temporary immobilization, 116
Terminal
   blackout time, 111-112
   control software, 54
   protocols, 102
   response, 131-132
Test
   considerations, 43
   coverage, 221
   development, 6, 23, 29-30, 196-200, 221-222
      design, 63
   environment simulator, 199
   feedback, 221
   mode, 95, 184
   patterns, 129
   phase, 7, 218-227
   quality and controls, 185
   reality, 222
   runs, 220
   technique, 220-221
   tools, 71, 196-197
   training, 223
Testing, 23, 63
   the documentation, 222
   the system, 184
   via one module, 197-198
Think time simulation, 199
Threat of automation, 100

Threshold costs, 160
Threshold limit checks, 109
Throughput estimation, 93-94
Time-current updating, 149
Timely information, 100
Tools, 23, 63
Traffic analysis, 161
Training, 6, 23, 24, 30, 63, 201-203, 223
   continuity, 179
   design, 63
   features, 178-180
   for distributed systems, 201
   messages, 70
   mode, 95, 174, 179
   of support staff, 223
   records, 180
   strategy, 201-203
Transaction
   analysis, 74
   compounding, 149
   counts, 74
   data element matrix, 80-82
   dispatching, 92
   driven systems, 91
   frequency
      analysis, 73-78, 88
      counts, 211
      diagram, 77, 151
      histograms, 77-78
   log file, 44
   processing time, 82-83, 233
   scheduling module, 44
   sequence
      analysis, 78
      diagram, 79, 151
      histograms, 79
      matrix, 211
   statistics, 45
   volume data, 74
Transition planning, 24, 63
Transmission rates, 181
Trouble indicators, 109
Troubleshooting data, 168
Tuning due to interference, 235
Turnaround time, 133

Unauthorized traffic, 126
Unique identifiers, 65, 81
Unit test, 23, 208
Units of equipment, 160
Unusual operating conditions, 107

Update synchronization, 93
Upline debug, 177
Usage and service, 122
User, 24, 63
   degraded mode information, 189
   management, 10, 24, 63
   organization, 53
   participation, 53
   rank, 45
   staff, 24, 219
   to manager messages, 237
   training, 201-202, 209
User's manual, 133, 200, 201
Using excess CPU cycles, 182
Utility programs, 116

Validation omitted, 190
Van Slyke, Richard, 117
Variable field length(s), 146, 147-148

Vendor personnal, 10, 37, 52, 60, 72, 104, 188, 206
Vendor response times, 105
Version numbers, 226
VIEW VERBS command, 140
Vocabulary controls, 59
Volatility ordering, 183

Warm start, 72-73
Weaning system operators, 230
Weekend operation, 121
Work area overlays, 170
Work in process conversion, 24, 208
Working space snapshots, 67, 170
Workload
   changes, 75, 116
   projection(s), 25, 32, 229
   statistics, 95

ZAP utility, 176